AUTOBIOGRAPHY
(1931–1967)

The author

AUTOBIOGRAPHY
(1931–1967)

Heir Apparent
Part One

Sadar-i-Riyasat
Part Two

KARAN SINGH

DELHI
OXFORD UNIVERSITY PRESS
BOMBAY CALCUTTA MADRAS
1989

Oxford University Press, Walton Street, Oxford OX2 6DP

NEW YORK TORONTO
DELHI BOMBAY CALCUTTA MADRAS KARACHI
PETALING JAYA SINGAPORE HONG KONG TOKYO
NAIROBI DAR ES SALAAM
MELBOURNE AUCKLAND
and associates in
BERLIN IBADAN

Heir Apparent: First published 1982
Sadar-i-Riyasat: First published 1985
Oxford India Paperbacks 1989

SBN 0 19 562436 X

Printed at Rekha Printers Pvt. Ltd., New Delhi 110020
and published by S. K. Mookerjee, Oxford University Press
YMCA Library Building, Jai Singh Road, New Delhi 110001

To Asha:

Enjoying love, children, fortune, wealth,
by being always behind thy husband
in his life's vocation,
gird thyself for immortality.

Atharva Veda XIV.I.42

and

To Vikram and Ajay:

Be not parted–growing old, taking thought,
thriving together, moving under a common bond,
speaking sweetly to one another,
have one aim and be of one mind

Atharva Veda III.30.5

Preface

To write about oneself involves, inevitably, a curious combination of egoism and introspection, specially if it covers both the inner and outer dimensions of life. I have been peculiarly fortunate not only in having been actively involved from a very early age in public affairs during a turbulent and fascinating period in our history, but also in having been exposed to a rich array of inner stimuli—intellectual, emotional and spiritual. These include direct contact with many of the leading figures in India's national life as well as with men of spiritual experience and wisdom. Also, through a love of books and music, I have travelled extensively in what Keats so appropriately called the 'realms of gold'—the luminous domain of ideas as represented in their highest watermark by the Upanishads in the Indian tradition and Plato in the Western; as also by the great treasures of Eastern and Western poetry and music.

Ever since my early youth I have felt that there has to be a deeper purpose in life than simply playing out passively the hand dealt to one by destiny or fate, chance or Karma—whichever term one may wish to use. Many of my political ideas and ambitions came to fruition in the period subsequent to that covered by this volume, but I also had deeper spiritual impulses which have remained with me and in fact are steadily deepening as I now enter the autumn of my life. In Kashmir, let me add, autumn is the most beautiful of all seasons. The startled efflorescence of spring and the steamy exuberance of summer give way to a maturation when the leaves turn into russet and gold, and the air is full of mellow sunlight.

For me, writing this autobiography has been a rewarding experience as it enabled me to look backwards in time and space, and inwards into myself. The material contained in this book was originally published in two separate volumes, the first twelve chapters as Heir Apparent in 1982 and the next twelve as Sadar-i-Riyasat in 1985. Though they were written three years apart, they form a continuing narrative and, between them, cover the period of my life during which I was predominantly involved in the affairs of my home state of Jammu & Kashmir. As the two volumes went into several editions, and also were not unduly long, my publishers felt that they could usefully be combined in the new Oxford India Paperback Series. This would be more convenient for the general reader, specially for students of modern Indian history and politics.

I am often asked if and when I am going to bring out the next volume. Curiously, my life seems to fall neatly into eighteen-year periods. The first eighteen were spent as a student in India and patient abroad, while no sooner had I turned eighteen than I found myself catapulted into the office of Head of State which I held under three different appellations—Regent, Sadar-i-Riyasat and Governor—for exactly eighteen years. Within a few days of my thirty-sixth birthday I was invited to join the Union Council of Ministers, and for the next eighteen years until 1985 was active in national politics as a Member of Parliament, including over ten years as Cabinet Minister. Obviously, if I do decide to continue the narrative, the next volume would have to cover those stimulating, challenging and often difficult years.

By the time I was fifty-four the overtly political phase of my life appeared to be behind me, but far from feeling any sense of deprivation I have found that I can now turn my energies more effectively to exploring the many dimensions of the human condition in this nuclear age, and in seeking to articulate a philosophy capable of dealing with the tremendous challenges that we face as we transit to a global society. For this one must look for the golden thread that runs through all the great spiritual traditions of the world, and that, by an astounding synchronicity, links with the latest insights of science—quantum mechanics and the new physics, extra-galactic cosmology and trans-personal psychology. This quest now takes me around the world, and offers me interaction with some of the most creative minds alive today.

Born as we are at a crucial juncture in human history, we must look upon life not as a meaningless meandering from the womb to the tomb, but as a breathtaking adventure in quest of a new consciousness, a new paradigm, a new pattern and structure of thought in harmony with the compulsions of the nuclear age. What the final upshot will be remains uncertain, but for me the path seems at this point to lie not in an ascetic withdrawal, but in a conscious effort to become a tiny but conscious element in the majestic rhythms of my chosen deity—Shiva Nataraja, Lord of the Cosmic

Dance. And this involves the search for a new dimension in which the outer and the inner life merge harmoniously into a luminous whole.

KARAN SINGH

9 March 1989

Contents

Heir Apparent

Chapter 1

The resort town of Cannes on the Mediterranean was always a favourite haunt of high society—European and, in other times, Indian. Among the several notable tourist establishments built along the famous Cote d'azur is the Hotel Martinez. Early in 1931 the dashing and handsome Maharaja of Jammu and Kashmir, Sir Hari Singh, and his lovely wife, Maharani Tara Devi, checked into the entire third floor of the hotel. The Maharaja was representing the Indian Princes at the Round Table Conference held in London that year. But London in the winter was foggy and unpleasant, while Cannes was glorious, with lots of polo and champagne and the balmy breeze from the Mediterranean so greatly prized by an aristocracy sublimely unaware that its days throughout the world were rapidly drawing to a close.

The Hotel Martinez still stands, a square, handsome structure commanding a fine view of the Mediterranean. In the northern corner of the third floor in suite 318–19–20 the young Maharani was in an advanced stage of pregnancy. She was only twenty-one, and was attended upon by a bevy of maid servants, some from India and others from France, for in those days servants in Europe were still in the realm of the possible. There were last minute complications, labour pains were unduly delayed, the maid servants, the Maharaja's personal physician Colonel J. H. Hugo and the eminent obstetrician Sir Henry Simson worked round the clock, while the Maharaja with his friends and staff was playing polo during the day and drinking champagne deep into the night. At last the appointed hour struck; on the ninth of March 1931 I came into the world this time round, nine pounds in weight and bawling vigorously. The odyssey had begun.

My birth was greeted with extravagant enthusiasm by the people of Jammu and Kashmir, irrespective of religion, caste or creed. Partly this was because the princely order in India still had some glory, and the birth of a Yuvaraj (heir apparent) was always a matter for rejoicing. But in my case the causes appear to have been deeper. My father had ascended the throne in 1925 on the passing away of his uncle, Maharaja Pratap Singh, who had ruled the State for forty years. During that time the British, alarmed at the growing power of the Russian Empire, had moved to consolidate their grip over the strategic northern provinces of Gilgit and Skardu. These were contiguous to the Russian Empire and had been added to the State of Jammu and Kashmir by the vision of Maharaja Gulab Singh and the military genius of his great generals, led by Wazir Zorawar Singh. Indeed, during Maharaja Pratap Singh's reign the British Political Department had worked up an elaborate intrigue to depose him on the grounds of alleged collusion with a 'foreign power', and had in fact obliged him to surrender power to a Council of Regency. He might have been actually deposed, but the plot was revealed by the *Amrita Bazar Patrika* of Calcutta in a celebrated article entitled 'Condemned Unheard'. This caused a furore in the British Parliament and the Political Office had to back down.

My father had married thrice earlier, twice in Saurashtra and once in the neighbouring hill State of Chamba, but his first wife had died with a child still in her womb, and the other two marriages were childless. People had begun to fear that if there was no heir to the throne, my father himself was an only child, the almighty Viceroy would one day invoke the notorious doctrine of lapse and take the State over under direct British administration. Thus when my father decided to marry once again—this time a girl from a remote village on the banks of the river Beas in Kangra district of the old Panjab—fresh hopes were raised. And when it was rumoured that the new Maharani was expecting a child, excitement began to mount.

My father's decision to take my mother to Europe for the delivery was variously interpreted. Some felt that he was doing this to keep the mother and child away from the malevolent machinations of the *zenana*, the female household where dozens of Maharanis, Ranis, ladies-in-waiting and maid servants lived —relics of previous rulers. Others assumed that France had been chosen because there the agents of the ubiquitous British Empire, upon whom at that time the sun never set, would not be able to operate. In all events, the birth of the Yuvaraj, if contemporary records and accounts are to be believed, triggered off an almost delirious wave of enthusiasm among the people of the State. There were official proclamations. The slaughter of animals, fishing and shooting were prohibited for three days, the 10th, 11th and 12th of March, and these were declared public holidays in the State: Offerings were made in temples, mosques and gurudwaras, and all school children were given sweets and asked to pray for the long life of the prince. I still meet people in the most unlikely places who remember having received the sweets as children. The announcement of my birth was made in Srinagar by the Army and Public Works Minister, General Janak Singh, while in Jammu this was done by Mr Wakefield, one of my father's Ministers.

General Janak Singh recorded a brief note on 17 March which contains the following paragraph:

> The 9th March was the first bright day when the weather in Kashmir cleared after a long spell of bad winter. It was on this day that the newspapers containing the happy news of the Gandhi–Irwin agreement reached Srinagar. The people of Kashmir thought these happenings a good omen and believed that the future destiny of the Prince was bright.

After spending six weeks in Cannes the whole party sailed back to India on the *Kaiser-i-Hind*, a P & O steamship, which docked in Bombay at the end of April 1931. It was thus Bombay, not Kashmir, that was my first point of contact with India and, strangely enough, Bombay was also to play an important part in my life for the next thirty years. A large group of officials and courtiers from Jammu and Kashmir had assembled at the Gateway of India to welcome my father, along with a number of ruling princes. Among the latter, as my father subsequently was fond of recounting, was the late Maharaja Ganga Singhji of Bikaner. Apparently I took one look at his famous moustache and set up a loud yell which subsided only when he finally left the cabin. There were also the late Nawab Taley Mohammad Khan of Palanpur, a close friend of my father, after whom our Srinagar house was named, and the 'Nightingale of India', Sarojini Naidu. From Bombay the party went by train to Jammu where on 3 May a triumphal reception awaited us. My father and mother went through the town in an open horse carriage, while I was driven behind in a car along with my English nurse Miss Doris Tranchell. Then, after five days, the whole performance was repeated in Srinagar. From the files and numerous people I have met who were present, it is clear that for days there was a staggering array of feasts, receptions, banquets, illuminations, free cinema shows, music performances, the distribution of sweets and sundry festivities. My formal naming ceremony was performed on 11 May, and Mr Wakefield solemnly announced that I would be called 'Shree Yuvaraj Karan Singhji Bahadur'.

Historically, this could perhaps be described as the zenith of Dogra rule in Jammu and Kashmir. Almost immediately after the festivities were concluded the State was plunged into serious political turmoil, after which things were never again to be the same. Oddly enough, my birth coincided with the emergence of a hitherto obscure schoolteacher into the political life of the State, with repercussions that continue down to the present day. His name—Sheikh Mohammed Abdullah. One theory is that the disturbances and the Sheikh's activities were in fact master-minded by the British, both because my birth had upset their 'doctrine of lapse' policy, and also to teach my father a lesson for having made at the Round Table Conference earlier that year a remarkably patriotic speech urging the British to respect the aspirations of the Indian people. He had said in the course of his speech that 'as Indians and loyal to the land whence they derived their birth and infant nurture, the Princes stood solidly as the rest of their countrymen for India's enjoyment of a position of honour and equality in the British Commonwealth of Nations'.

Be that as it may, from as far back as I can remember, I have been regaled with glowing accounts of the festivities and celebrations that followed my birth, coupled occasionally with unusually mysterious astrological predictions regarding my future life. After a while I began to wonder whether what I did with my life would in fact justify the joy and expectations that my arrival appeared to have elicited. If even a fraction of the people who rejoiced had done so from feelings deeper than formal loyalty, it was a debt that would have to be repaid in full measure in the years and decades ahead.

Looking through the yellowing files I find that all the religious ceremonies connected with my childhood were performed with the meticulous attention to detail that was characteristic of my father. Programmes were neatly printed, and every feature clearly spelt out down to the minutest item. The *Annaprashan* (cereal tasting) ceremony was held on 8 February 1932 and the *Mundan* (tonsure) on 7 December 1933. Every year until 1947 my birthday was treated as a public holiday; a salute of 17 guns would be fired from all forts; shooting, fishing and the killing of animals were prohibited; prisoners were released; alms distributed to the poor and all palaces and public buildings in Srinagar and Jammu illuminated. Religious ceremonies were held to which I was taken, attended by my father, mother and other relatives.

My earliest recollections are of looking up at the sky outside Amar Mahal, our mini-castle in Jammu which is now a museum and library, with someone pointing out an eagle to me, a tiny black speck against the unending blue vastness; and later, a curiously vivid memory of looking at myself in a mirror in a smaller building within the same campus, almost as if I was seeing myself for the first time. At the age of three I was taken away from my mother and an independent establishment set up for me in separate houses, at Srinagar in summer and Jammu in winter. I was allowed to see my mother every day for only one hour, and my father thrice a week. This was obviously not an ideal family environment, and it flowed from a deep incompatibility between my parents. My mother was a village girl from Kangra; my father was the ruler of the largest of India's over five hundred native States. My mother was deeply religious; my father remained till the end of his days a virtual agnostic. My mother was warm, gregarious and loved children; my father was stern, severe and moved only in a carefully chosen circle of courtiers and very few friends. My mother was strong on conversation; my father was so formidable that normal conversation in his presence was virtually impossible; my mother was superstitious, demonstrative, emotional; my father was neat, meticulous, aloof. This psychological and emotional imbalance led to a good deal of tension and mutual conflict.

Understandably, my early sympathy lay almost entirely with my mother. I cherished the visits to her, while she doted on me and would count the hours for me to appear. She resented my having been taken away from her on the grounds that she would spoil me, and was often in tears when my allotted hour was over. For many years the knowledge of her distress made me deeply unhappy, and I would often lie awake at night crying softly, thinking of her. She was beautiful, with large, expressive eyes which, as a friend of my father used to say, she knew how to use. To me, of course, she was for years the epitome of grace and love. When I was brought to her she would first take me to her prayer-room where she would put flowers and a few coins into my hands, to be offered to the various pictures of gods and goddesses installed there. Then we would sit and play games with her maid servants, or with her nieces and nephews, my only relatives of comparable age, as there were no close relations on my father's side of the family.

Sometimes in the mellow Jammu evenings we would all assemble on the large veranda in Amar Mahal with its magnificent view of the Shivalik range, the triple-peaked mountains of Vaishno Devi dominating the skyline and the river Tawi winding down to the plains. There, the small earthen lamps were lit, and with my mother in the lead we would all walk round the sacred *tulasi* (basil) plant grown in an earthen pot, singing devotional songs in our lilting mother tongue, Dogri. Years later I translated one of those songs dedicated to the great Mother of Radiances:

O Mother Jwala, dwelling amidst the mountains,
fulfil our innermost desires.
A bright red garment adorns your body
and on your forehead is the yellow saffron mark,
the five-hued shawl covers your head,
its edges shimmering with golden embroidery,
O Mother Jwala, dwelling amidst the mountains,
fulfil our innermost desires.

From all corners of the earth, O Mother,
pilgrims come and sing thy praises,
having bowed before thy shrine
all their cravings are satisfied,
O Mother Jwala, dwelling amidst the mountains,
fulfil our innermost desires.

Brahma, the Creator, recites the Vedas before thee
and Shankar meditates upon thee amidst the mountains;

the devotee who sings thy praises
is granted by thee his heart's desire,
O Mother Jwala, dwelling amidst the mountains,
fulfil our innermost desires.

Although I did not then understand the full meaning of the words, the rhythmic mingling of the sacred and the mundane, of the here and the hereafter, had a deep effect upon me. My mother loved folk music. She had a powerful voice and would sing for hours with the *dholki* (two-faced drum), in chorus with her maids and other female relatives and ladies from the city who visited her. Evidently I developed my permanent romance with music from these early childhood memories of Dogra–Pahari songs, because for the rest of my life music has been a predominant motif in my consciousness.

While the daily, and later only thrice-weekly, visits to my mother were eagerly awaited, the meetings with my father were a source of awe. He was affectionate enough, without being demonstrative, and always carried my photograph in his gold cigarette case. But his whole presence and reputation were so formidable that it was difficult to think of anything to say. The joyful spontaneity of the visits to my mother was lacking, and it was only many years later that I began to realize that my father's forbidding exterior was really something of a protective armour that he had developed through the circumstances of his own life. An only child, brought up in a cloak and dagger atmosphere of court and family intrigue, he must have been through a traumatic situation before he grew to manhood. And soon thereafter, on his first visit to England, he became the unfortunate victim of a vicious blackmail plot that brought him a great deal of undeserved censure. By the time I was born my father was thirty-six and had been on the throne for six years. He always took a close and careful interest in the way I was brought up, but being by nature undemonstrative his relationship with me never became as free as my relationship with my mother. There was a brief period when it might have flowered, but destiny and the inexorable thrust of history intervened to thwart this.

Chapter 2

Along with a separate establishment consisting of about fifteen employees, I was looked after by a series of British guardians until I went to public school at the age of eleven. The earliest was a Mrs Wyndham, whose husband, Colonel Wyndham, was Assistant Resident. I remember her only as an elderly woman with a broad hat, and recall the Colonel once giving me an empty metal cigar box lined with crackling aluminium foil, silver on one side and bright red on the other. After the Wyndhams came the Ritchies. He had been a master in Aitchison Chiefs College, Lahore and was evidently Scottish, because my only clear recollection is of his teaching us some charming Scottish folk tunes and marching songs. Mrs Ritchie was a bit of a dragon whom all of us, including her husband, treated with rather nervous respect.

At about this time, around 1935 when I was four, two companions came to live with me, sons of Rajput courtiers close to my father. They were Digvijay Singh ('Digby'), who was the son of Rao Bahadur Kartar Singh, a senior and shrewd administrator who played an important role in the administration of the State; and Digpal Singh ('Bilti'), whose father, Major Faqir Singh, was for many years my father's favourite staff officer. When they came to live with me I was repeatedly told of my father's rather un-Christian instructions that if I ever slapped them once they should respond by slapping me twice. As it turned out the admonition was unnecessary; they were both older and stronger than I and the occasion to implement the instructions never arose.

By then I had acquired the nickname 'Tiger'. Apparently one day I crawled into the room where my father and his close friend, the ruler of Jodhpur, Maharaja Umaid Singhji, were seated. Umaid Singhji said that I was on all fours like a tiger, and the name stuck. My father and his friends invariably referred to me as Tiger, and years later Jawaharlal Nehru and others of his generation did the same. It is an odd coincidence that decades later I was called upon to head Project Tiger, designed to save this magnificent animal from extinction, and also that as Chairman of the Indian Board for Wildlife I was responsible for making the tiger India's national animal in place of the lion.

The three of us—Tiger, Digby and Bilti—lived together for several years in a world of our own. In Srinagar we were at the lovely cottage on Shankaracharya Hill, then known as Taley Manzil after Taley Mohammad Khan of Palanpur, a close friend of my father, which commands a fantastic double view. On one side is the Dal Lake behind which towers a range of high, bare mountains with the peak of Harmukh, eternally capped with snow, peeping over their shoulders; while on the other side the entire sweep of the snowy Pir Panjal range stretches from Gulmarg in the east to Banihal in the west. In Jammu we lived in a small building known as Karan Niwas, originally a staff house attached to one of the larger palaces in the estate. There was a full complement of servants with whom we played football, while my father's Goan butlers (Fernandez, Lobo, D'Souza) came down occasionally to play badminton. We had a strict regimen, dietary, academic and recreational. We were not allowed any Indian food or sweets, and therefore had to make do with the English food which the guardian of the moment and his wife relished. Tea was our best meal, for which there were buttered scones and sponge fingers, delicious shortbread made by Huntley and Palmers, and the inevitable sandwiches served on silver dishes. Indeed, breakfast and tea were the only meals in which the English culinary genius (such as it was) seemed capable of any sort of coherent expression.

For our studies we had an Indian tutor, Amar Nath Khosla, who taught us English, mathematics and other subjects. He had a neat handwriting, and every month would prepare a meticulous report on the three of us, covering our studies, games and general behaviour, which went to the guardian and, through him, to my father. It was Masterji, as we called him, who one day brought us a black marble tablet with the figure of a frail, dhoti-clad man inlaid in mother-of-pearl. This, he said, was Gandhiji, the great leader of India. But he only showed it to us and did not leave it with us for fear that he would get into serious trouble with the guardian. We were also initiated into the evocative if somewhat inelegant gibe that if all Indians were to urinate together it would be enough to wash all the Britishers in India into the sea. For some reason the three of us were anti-British, though we never showed this openly. I distinctly remember listening to the radio in September 1939, when I was eight and a half years old, and hearing the news that the Second World War had broken out. Thereafter we were all secretly rooting for the Germans, and would jump around gleefully whenever there was news of British reverses. Only one servant, Raghunath Singh, would infuriate us by insisting that the British would never be defeated despite initial setbacks. We would all taunt him, but he stuck imperturbably to his guns, Churchillian in his supreme confidence of ultimate victory. I discovered later that he had been for some time in one of the regiments of the Indian Army.

The Ritchies were in due course succeeded by Captain and Mrs Wreford, who turned out to be our favourites among the British. Captain (Reggie) Wreford, formerly Census Commissioner in the State, had lost an arm in the First World War, but he used his stump to astonishing effect and was an excellent shot. Katherine Wreford was a woman of charm and quiet efficiency and their children, Joan and Douglas, both considerably older than us, were at school in England and came out every year on their holidays. During the Wreford tenure we started learning badminton, tennis, football and hockey. We were also initiated into cricket, and would write letters to Bradman and Hammond and other such legendary heroes. Despite this, cricket did not become one of my favourite games, and I have never had more than a desultory interest in this curiously protracted sport. Occasionally we were allowed to go shopping, usually in Gulmarg which at that time had glittering shops filled with the latest foreign products to cater to the large number of British people who flocked there in the summer. During the winter, when in Jammu, our shopping consisted of regular fortnightly trips with our guardian to Sialkot, only an hour's drive away. There we would have tea at Ghulam Qadir's, a department store, and buy comics from a small corner bookshop opposite. We also subscribed to three comics from England—*Tiger Tim* (mine), *Donald Duck* (Digby's) and *Puck* (Bilti's).

My father was an excellent rider and, while he was too heavy to be really top-class, he reached the respectable handicap of six in polo. He was keen that I should become a great polo player, and from the age of three I was forced to ride almost every day of my life, beginning on a box saddle and a tiny horse and gradually moving on to larger animals. I had several falls, and on one terrifying occasion my horse bolted in the Jammu pologround and I hung on to it, petrified, until it finally exhausted itself and came to a halt. During those years I saw some of the finest polo in the world as the famed Jaipur team with the remarkable combined handicap of 35 (out of a possible maximum of 40) would come up to Srinagar every year. We would watch the game from the special Zenana pavilion along with my mother, my grandmother and other elderly ladies in white. These were among the few occasions when I met our elderly female relatives, though every two or three months my mother would take me to the Zenana palaces to visit them.

An event I remember was the great Quetta earthquake in May 1935. For weeks we were not allowed to spend the night inside Taley Manzil and, to our great excitement, had to sleep out in the garden. Also, there was the occasion when my tonsils had to be removed. A room in Karan Niwas at Jammu was converted into an operating theatre, and the well-known surgeon, Colonel Harper Nelson, came over from Lahore to do the operation. I remember struggling with horror under the chloroform mask held over my face and, later, my father paying one of his rare visits to me and sitting on my bed with a jigsaw puzzle, pleasantly surprised that I was eating ice cream because it was supposed to be good for my post-operative throat.

Apart from the fact that my first contact with Indian soil took place in Bombay, that great city became an important venue for my early experiences. Though in his early life an avid polo player, my father later took to racing in a big way and this inevitably led him to Bombay which had the best racing in India. Consequently, while he spent the six summer months in Srinagar, in the winters he would spend only about two months in Jammu and four in Bombay. The first time we went to Bombay, in 1940, he stayed in a rented building called Nishat House on Carmichael Road. All I can recall of that visit is that we used to get a delicious orange drink called Pure Gold, sipped luxuriously through straws, and also that the day before we left the three of us were roundly reprimanded for having ruined our sitting-room by spilling ink all over the sofas and carpet. The next year my father shifted to a block of flats he bought at 94 Nepean Sea Road. He lived on the top floor, the sixth, and for me he kept the first floor which had a terrace garden—a novelty in those days.

We lived there for some months with our guardian. By then the British guardians had been dispensed with, and Colonel Kailash Narayan Haksar appointed in their place. It was an open secret that my father wanted him as political adviser, and adopted the somewhat transparent stratagem of appointing him guardian to the Heir Apparent. This highlighted a strange tendency in my father that was ultimately to prove his political undoing—his inability to trust anyone for any length of time. He selected his Prime Ministers with great care, but no sooner had he appointed one than he would begin cultivating someone else as a sort of counterbalance. With the exception of N. Gopalaswami Ayyangar, who was Prime Minister of the State for six years from 1937 onwards, this happened with all the succeeding incumbents—Raja Maharaj Singh; B. N. Rau; Colonel K. N. Haksar; Pandit Ramchandra Kak and General Janak Singh—right down to the crisis of 1947.

Colonel Haksar was a large man with bushy eyebrows, gold-rimmed spectacles and a military limp which combined to give him a formidable appearance. But despite his stern exterior he was fond of me, and would often seat me on his knee with great affection. My father, I suspect, was a trifle scared of him too, and though he joked about him with us in his absence, he was discreet when 'Colonel Sahib' was in the room. Colonel Haksar, in turn, while loyal to my father, by no means approved of all that he did. I remember him telling me one day with more than a touch of asperity as we drove away from a meeting with my father, 'Tiger, look very carefully, so that when you grow up you will know what *not* to do.' Colonel Haksar's eldest daughter Shyama Wattal had a marvellous way with children, and at her coaxing I wrote my first 'book', an eight-page, handwritten collection of poems and mini-essays.

It was in Bombay in 1940 that I was sent to my first school—the Cathedral High School—along with Digby and Bilti. At that time the school catered predominantly to English, Parsi and Anglo-Indian boys, and we were known as Singh I, Singh II and Singh III. These peculiar appellations became a bit of a bore, and although I went back for one more summer term the next year, I cannot say that I really derived any substantial benefit from that estimable institution. Piloo Mody, who was also at the Doon School when I went there later, has recorded that Zulfikar Ali Bhutto was his classmate at Cathedral, which means that we must have been there at the same time, although I do not remember Bhutto at school. The teachers were all English, even the Urdu ('Oordoo') teacher, who didn't seem to be able to speak a coherent sentence of that beautiful language himself.

Apart from some fellow princes, my father's friends were drawn largely from racing circles, but there was a small group of foreigners whom I remember vividly. Victor Rosenthal, a French jeweller who had enjoyed a fabulously chequered and romantic career, making and losing many fortunes *en route*, was a fascinating man with a chiselled beard and a warm temperament. He had known my father for decades, and handled much of his jewellery and foreign investments. He was a regular visitor to Kashmir in the summers, and always brought me lovely presents—books by Maupassant and O. Henry which I devoured with relish, a set of plastic hairbrushes, boxes of Swiss chocolates. I have described him as French simply because he lived in Paris and we all thought that was his nationality. It was only many years later that I discovered he was, in fact, Russian, having moved out—complete with beard—at roughly the same time as Lenin moved in. Freddie and Beryl Stileman, an English couple, were good friends of my parents. He was in the British firm of Killick Nixons, while Beryl was a woman of great beauty and vivacity. Almost exactly my father's age, she was the only person I remember who could tease him and speak quite naturally to him, a feat which in those days was looked upon as bordering on the supernatural.

Among the Indians, M. H. Ahmedbhoy and his wife Rukhsana were frequent visitors, not only because of his racing background but also because he was a superb singer. Despite exaggerated gestures as he sang, 'Mahmood Seth' was one of the finest amateurs I have ever heard. *Thumries* were his forte, and along with him my father frequently organized music parties where some of the best artists in India would perform, among them Kesarbai Kerkar, Siddheshwari Bai, Begum Akhtar and Meneka Shirodkar. My father did not sing himself, but he had a fine ear for good singing. It was at about this time that he got me to start learning classical music, an act for which I will remain eternally indebted to him. My first teacher was Balram Singh Rawat from the Nepal Terai. No great musician, he was a competent teacher and within a couple of years I had picked up several ragas. Ustad Vilayat Husain of the famous Agra gharana came to coach me for two summers. About three years later my father had HMV make some records (78 rpm) of my songs for private circulation, which he proudly sent round to his friends.

Music has remained one of the major forces in my life. The very first tune I can remember goes back to when I was four or five; a popular devotional ditty in praise of the boy-God Krishna, 'Sunā de, sunā de, sunā de Krishna, tū bānsuri dī tāan sunā de Krishna'. Some years later a new cinema hall opened in Srinagar called the Amrish Theatre (the owner, Pandit Krishan Bal, had wanted to name it after me but my father refused him permission. He therefore named it after his youngest son). I went with my parents to the opening, for which the celebrated dancer and actress Sadhona Bose had come all the way from Calcutta. As far as I can recall, her dancing on that occasion consisted largely of striking a variety of sculpturesque poses, scantily attired, upon a huge drum. However, there was another dance in which a man dressed in blue (the ocean) and a girl in green (a river) did a charming little number. The tune is one that will remain with me as long as I live. It was in the pentatonic raga, Bhupali. As soon as I heard that tune it was as if in some curious way a fountain had been released within me and the music seemed to pour out of my finger tips. For weeks the flow of those fine notes coursed through my veins. Later I learnt from Balram Singh several songs in this raga, as also in two other ragas equally wondrous—Durga and Malkauns. Once the seed of a new tune was planted in me it would germinate and grow, until I would find my whole being pulsating to its rhythm.

My father was a crack shot, both with rifle and shot-gun, and held international records in shooting duck and the Kashmiri partridge, the *chakor*. His shooting parties were meticulously planned, each guest provided with cartridges and a packed lunch complete with red wine. Duck shoots at Hokarsar and Higam, shallow lakes to the west of Srinagar, would begin early in the mornings. We would all drive out in a great procession, assemble on the banks of the lake, and at the appointed hour get into our boats and row to our respective 'butts'. There the whole day would be spent massacring duck of various species, including occasional wild geese. As a special treat my father would sometimes let me sit behind him while he was shooting, and I would click the gold counter whenever he dropped a bird. He would use three double-barrelled Webley & Scott shot-guns, with a loader standing behind him on either side. After firing the first gun he would hand it over to the man at his left and take a loaded one from the one on his right. By the time he had fired the next gun the previous one had been reloaded, and this relay would carry on for hours. I recall several occasions when a flight of duck came across the lake and my father dropped two birds with the first gun when the flight was just within range, another two with the second when it was overhead, and yet another two with the third when it had passed him but was still within range. I admired his skill tremendously.

At big game, also, he was superb. In the sub-valleys around Kashmir, and in the vicinity of Udhampur in the Jammu region, he had developed some excellent shooting reserves. I was with him on several occasions when he shot black bear, wild boar, panther and—alas—the magnificent Kashmir stag (the Hangul), which is now almost extinct. The air in the cold October mornings was crisp and clear, the hillsides ablaze in russet and gold, and the atavistic lure of the hunt caused the blood to flow faster through one's veins. Apart from shooting, the trout fishing in Kashmir is among the best in the world. A century ago my great-grandfather Maharaja Ranbir Singh had sent Queen Victoria a present of a pair of Kashmir pashmina goats. The man who came to get them reported back to Her Majesty that the mountain streams of Kashmir were ideally suited to trout, whereupon the redoubtable monarch sent back a return gift of a tank full of tiny trout fingerlings.

These flourished so well in Kashmir that they soon far outgrew their ancestors. In England a fish over a pound is considered a good catch; I have known days when my father would throw back anything under two. His largest catch was 14½ pounds, but in the Harwan hatchery a brown trout grew to an astonishing 27 pounds.

Accompanying my father on a fishing trip to Thricker and Nambal on the Liddar river coming down from Pahalgam was a cherished event. The stream flowed literally through the garden, and one would get up early and catch fish for breakfast. The whole day we would be out fishing, and in the evenings return to the lodge. If my father landed a good catch there was merriment; but if he had a bad day or lost a big fish he would be in a terrible mood. This almost invariably meant that some unfortunate servant or shikari was sacked in the middle of the night. All of us then sat quietly, listening to the sound of the petromax lamps and the stream flowing outside, not daring to speak.

In fact, father shared a trait common to the feudal order; he was a bad loser. Any small setback in shooting or fishing, polo or racing, would throw him into a dark mood which lasted for days. And this would inevitably lead to what came to be known as a 'muquaddama', a long inquiry into the alleged inefficiency or misbehaviour of some hapless young member of the staff or a servant. As I grew older these cruel inquisitions began to infuriate me and added to my growing alienation from the order into which I had been born. Here was authority without generosity; power without compassion.

Dachigam, just twelve miles from Srinagar, was one of my father's favourite haunts. Beyond the reservoir at Harwan where, twenty-four centuries ago, the second international Buddhist conclave is believed to have been held, the road turns into a steep valley. Four miles from the entrance to the reserve there is a clearing in which my father's famous shooting lodge was constructed. The game reserve teemed with black bear and wild boar, while in the autumn the magnificent Hangul would move down, along with the descending snowline. The lodge was simple but elegant, consisting of six bedrooms with attached baths, and a drawing and dining-room, the walls covered with beige silk. It commanded a spectacular view; heavily afforested mountains soared into the sky on three sides, while just behind the lodge a lower hill covered with grass and dotted with black boulders provided the perfect counterpoint. The garden had four large walnut trees, beds of multi-coloured roses, and lots of green lawn.

We came here at least half a dozen times in the season. My father would occasionally shoot a bear or pig—once he shot a bear from the veranda of the lodge—but more often there would just be a picnic. His shikari was a gnarled, sun-beaten Kashmiri called Rahman Wani, with a face like carved granite and impressive red moustaches. We came sometimes with my mother, her lady-in-waiting Wazirni Sita, and a wide assortment of relatives and maid servants, and played hide and seek in the compound among the rose bushes. The singing of the wind through the great trees, the softly insistent gushing of the stream that flowed through the valley, the blue of the sky and the green of the forest always in harmonious contrast, combined to give Dachigam a magical quality. One entered there an almost different dimension, a world where human beings were in a way redundant and the elements held their gracious sway over time and space.

Another marvellous spot was our cottage on an island in the Dal Lake which was originally my grandfather's pigeon house. This led it to be called 'Kotar Khana' in Kashmiri, and although it was subsequently renamed Lake Pavilion and then Lakshmi Kutir, the original name has stuck. The cottage commands a truly breathtaking view of the mountains spread out a hundred and eighty degrees like a gigantic amphitheatre. I have always loved being among mountains, though I have never been possessed of the desire to climb them. I like just looking at them, feeling them as exhilarating presences in my consciousness. Their poise, unhurried and imperturbable, provides a welcome contrast to the rush and bustle, the envy and intrigue, of daily life. I have had occasion to travel to the ends of the earth, and have found that all great mountain ranges—the Himalaya, the Alps, the Andes—share this very special quality of expanding human consciousness.

It was the stunning natural beauty of Kashmir that first taught me to ask deeper questions leading into the heart of the mystery that is life. In my early years my religious feelings were largely confined to the traditional devotionalism of my mother—genuine but somewhat limited. Looking back, though, it is clear that the sheer physical beauty of Kashmir—its mountains and valleys, rivers and forests, paddy fields and winding mountain tracks—helped to sharpen my awareness of beauty, to make me more responsive to the softer and finer aspects of the human spirit that are so often steam-rollered into the dust by the hammering insistence of the life of the moment. Growing up in beautiful natural surroundings is a rare and inestimable privilege. It is curious how in India we tend to ignore natural beauty, so that children often remain quite unaware of this dimension. In fact, it was my British guardians, and people like Beryl Stileman, who constantly went into raptures over lovely views and sunsets. At the time this rather amused us, and we felt secretly superior on the assumption that India was evidently much more beautiful than England. But gradually I began to see our natural beauty with new, more open, eyes.

Social awareness does not come easily to a prince. I lived in a self-contained world where servants were taken for granted and the question of why some should be the rulers and others the ruled was never formulated, far less asked. And yet I had one advantage over others in my position. My mother was a village girl from a poor family, and howsoever much she may have liked the trappings and tinsel of royalty, she always considered it a sacred duty to alleviate the want and suffering of the poor. Throughout her thirty years as Maharani she spent large sums on helping not only poor relatives, but hundreds of common folk in want and distress. It is impossible to recount the number of girls she got married off, the houses for the poor she had made, the clothes and sweets she would forever be distributing. She always brought home to me the fact that I had a duty to serve and help the poor. 'If you help the rich', she would say, 'they will only take your money and turn upon you when you stop giving. Help the poor, they will appreciate it and pray for you and bless you.' In a way this was my introduction, if not to socialism, at least to the concept of distributive justice. She would also enjoin upon me to acknowledge every salutation with folded hands, to inquire after the welfare of people's families, to mix with whosoever came, rich or poor. 'Your father never meets the people,' she would complain, 'that's the trouble. He just sits surrounded by fawning courtiers and favourites, and never really gets to know what is going on outside.'

Apart from a series of ADCs drawn from the Jammu and Kashmir State forces, both Hindu and Muslim, my father had a number of prominent Muslim courtiers. Chief among them was Nawab Khushru Jung, a nobleman from Hyderabad whose services had been loaned to my father by the Nizam. 'Mahboob' was a close friend as well as Military Secretary to my father for many years. He had an impressive presence with his rich voice and courtly manner, was an excellent horseman and also, as even we boys knew, a great ladies' man. Sahibzada Nur Mohammed Khan, a portly figure from Baluchistan, was on my father's staff for several years. Then there was Sardar Abdul Rahman Effendi—'Bhaijan'—who was a close friend of my father's and lived on Gupkar Road just outside our Srinagar palace gates. He was a refugee from Afghanistan, being related to King Amanullah who had been deposed and forced to flee the country. Effendi's second wife was a large woman with an amusing Pushto-Urdu accent; she played tennis with my mother, while Effendi's mother-in-law was a memorable personality known as Koko Jan. When I saw her she was almost eighty but still always the life of the party. Apparently she had played a prominent role in the Afghan court during her youth. There was never any differentiation between the Hindu and Muslim courtiers, and, if anything, the latter seemed to be more favourably treated. As far as office work was concerned, this was mainly managed by a diminutive but highly efficient Kashmiri Pandit, Dina Nath Jalali, who had been with my father for an incredible twenty-five years, and a taller associate, Shambhu Nath Wanchu.

Kashmir then, as now, was a tourist's delight, and a galaxy of Indian princes and other personages would come up every summer, some as guests of my father. I remember the strikingly beautiful Princess Niloufer, then unhappy in her marriage with the Nizam of Hyderabad's second son, and her less attractive cousin Princess Durisheshwar, wife of the Nizam's elder son the Prince of Berar. A whole spectrum of Maharajas, Rajas and Nawabs were around. Specially close was the Nawab of Palanpur and his beautiful Australian Begum. I would meet some of these princes and would have to touch their feet and call them 'Uncle'. On one such occasion I was fascinated to meet a man who was stone deaf, and with whom conversation had to be carried on by means of a writing tablet. He was the Maharaja of Jind, famous for his extensive kennel. It struck me even then that the money he lavished on his dogs could with greater advantage have been spent on the welfare of his people.

The British were much less in evidence. Unlike British India, where their presence was ubiquitous, in the Indian States they were generally not too visible. This was specially so in our State, because my father genuinely distrusted them and even persuaded them to move their residency in winter to Sialkot in the Panjab rather than to Jammu. For the summer months the British Resident lived on the bund in Srinagar in the house later converted into the Government Handicrafts Emporium, but we seldom saw him and I cannot remember a single occasion on which I met him at the Palace. He held a fancy dress party every year, however, which I would attend along with Digby and Bilti. The Resident's wife would call on my mother occasionally along with other foreign ladies and the wives of non-Kashmiri officials. Among these I recall Lady Dalal, wife of the Parsi Chief Justice, Sir Barjor Dalal, a gentle little woman who would give me tiny pencils whenever she came and which I collected avidly.

As for the Kashmiri Muslims, our contacts were mostly limited to the gardeners and the shooting and fishing guards. Once my father asked Ghulam Ahmed, a jeweller and carpet manufacturer of repute who was also his art adviser, to take me round the city. This he did, and I remember my astonishment at seeing all those dilapidated buildings on the Jhelum, looking as if they could topple any moment into the river. 'These are your people', Ahmed said somewhat dramatically. The visit had a rather disturbing effect upon me, and for days thereafter the recollection of the Jhelum dwellers would flash back upon me, their squalor in such glaring contrast to the orderly elegance of the palace as to constitute virtually a different world.

Of political developments in the State at the time, or the great freedom movement sweeping the nation, we knew little. The *Tribune* and the *Civil & Military Gazette*, both published from Lahore, came to the house, but I was not yet ten and was too young to be able to read or understand them. The haunting signature tune of All India Radio (happily unchanged to this day) is among my earlier recollections, and we occasionally listened in to the news, specially after the War had begun. But we had no real idea of what was happening, and lived in an isolated world far removed from the economic and political realities outside. Despite our comfortable surroundings, I can seldom recall being really happy as a child. There was somewhere deep within me a built-in fear, undefined but nevertheless clearly felt. Perhaps it was because of the unsatisfactory parental situation, or my enforced absence from my mother. Whatever it was, it vitiated a good deal of my early life.

I have mentioned how I was sent to Cathedral High School in Bombay in the winter of 1940 and 1941. In the summer of those years I started attending the Presentation Convent School in Raj Bagh, Srinagar. In order to avoid travelling through the city over Amira Kadal, the first of the seven historic bridges across the Jhelum, I had to drive to the river bank near Pestonjees and Ahdoos (shops that had virtually become institutions), cross the river in a *shikara* and alight on the opposite bank quite near the school. The car meanwhile was driven across the bridge and awaited me on the other side. Thinking back, it seems that, because the National Conference movement against Dogra rule led by Sheikh Abdullah had begun gaining momentum, my father thought it safer for me to avoid driving through the city, although this was never given as a reason. The Presentation Convent was clean and efficiently run by an estimable group of Irish nuns led by Mother Peter. Our class teacher was Sister Annunciata, while the genial head of the mission, Father Shanks, would visit the school from time to time. They were all Irish, and spoke with a distinctive accent. Apart from the studies at which (alas, I suspect, not entirely on merit) I invariably stood first, there were music and games and plays and many other pleasant activities. Lovely pictures of the Virgin Mary and Jesus hung on the walls, and the nuns were models of courteous and genuine solicitude. I remember my first appearance in a play—it was about Robin Hood, and we worked on it for days before the annual function when it was performed. I was first cast as King Richard, but at the last moment my role was changed to that of Robin Hood. I believe my father felt that, being a prince, I should not assume that I would always have to play the royal role.

My father, however, was modern only in some ways. I have mentioned that he and the then Maharaja of Jodhpur were for years close friends. Every year Umaid Singhji and his strong-willed wife would come up to Kashmir with their children—five boys and a girl called Susan. I suspect that there had been some talk of Susan and me getting engaged, because that is the only explanation for the peculiar events that followed. For some reason which I have never been able to discover, the friendship between the two Maharajas broke. My father, true to the feudal tradition, was an extremist in such matters. For him there was no area of grey; a person was either in or out. And so, as a sort of reaction just to spite the Jodhpurs, he rushed into getting me engaged to another Rathaur princess, the daughter of the then ruler of Ratlam, Maharaja Sajjan Singhji. It will leave modern readers aghast that there could be serious engagements at the age of ten plus, but this was in the early forties and in an Indian State, and the betrothal of the Yuvaraj was a great ceremonial event. There was a full-dress darbar in the lawns of Gulab Bhavan, described by Reuters as 'one of those splendid displays of colour and magnificence for which Darbars in India are famous'. I was dressed in brocade and jewels, and have a clear recollection of walking into the gathering feeling absolutely ridiculous. A Ratlam courtier, Vijay Bahadur, came at the head of their contingent. He dropped a sovereign just before he was to present it to me, and whispered to me to pretend to take one off his empty handkerchief so that no one would notice his lapse. This I solemnly did as he was a good sport and I did not want to let him down.

The Ratlam princess Chandra Kunwer, called 'Shanti', and her younger brother came up to Srinagar with their guardian, a Mrs Stewart, and spent some months there. She seemed very nice but I had not reached the age of taking an intelligent interest in girls. Some years later, when I was in America, we started corresponding with each other. But the situation was obviously absurd, and the irony of the charade was compounded when, as I will record later, my father in 1949 equally impulsively broke off the engagement. I had no say on either occasion, but it is clear that the matter was handled in a manner that can only be described as disgraceful, grossly unfair to the Ratlam family—and particularly Shanti—and did no credit to ours. I learnt that Shanti subsequently married someone in Uttar Pradesh and many years later was injured in an automobile crash which ultimately led to her death.

Chapter 3

One of Colonel Haksar's grandsons, Vivek Nehru, was at the Doon School in Dehra Dun, which had been set up in 1935 on the pattern of the famous British public schools, the first of its kind in India. Colonel Haksar persuaded my father to send me there. Before this the Indian princes had either gone to school in England or to what were known as the 'princely' schools in Ajmer, Rajkot, Indore and other locations. Here the sons of the Indian aristocracy were educated, usually with the facility of private servants and lots of spending money. Sending me to the Doon School was, therefore, an imaginative and forward-looking decision which had a lasting influence upon my future life. Had I not been obliged to go through four rather unpleasant years at Doon, it would have been much more difficult for me to make the crucial transition from feudal to democratic life which lay only a few years ahead. At the time, however, the prospect was not a particularly appealing one. My mother wept copiously when I left, and ended fainting in the arms of her ladies-in-waiting. I was full of apprehension at the unknown life that awaited me at school where, as Colonel Haksar had repeatedly pointed out, I would have to make my own bed, polish my own shoes and be content with a pocket money of five rupees a month. Having spent my first eleven years in a sheltered and highly privileged situation, the change was obviously not going to be pleasant. Colonel Haksar himself escorted me to Dehra Dun, and in September 1942 I joined the school. I was allotted, expectedly, to Kashmir House and given the roll number 259.

My first few terms at school were miserable. I was desperately homesick, and wept as unobtrusively as I could every night before going to sleep. There were about a dozen rooms on each floor, four boys to a room, with two lavatories on each of the two floors of the house. I was never comfortable in the dark, and hated the damp bed sheets, the coarse blue blankets and the low slung mosquito nets. Luckily, I was never physically ill-treated by the other boys, but the fact that I was a prince was not something which particularly endeared me to them. A major difficulty arose because, largely due to the intensive coaching I was given in the months at home preceding my admission, I was placed two classes higher than usual for boys of my age. Although this was hailed by my parents as recognition of my outstanding intellectual capacities, it put me in a position where almost all the boys in my class were two years older than I, while those of my age were two classes below. In the result, I fitted in neither with my classmates nor with my age group. Added to this was the fact that I was rather timid and introverted, totally unprepared by my upbringing for the rough and tumble of public school life.

The food was virtually inedible, consisting largely of vegetables that I particularly loathed, such as turnips, cabbage and lady's fingers. The cumulative effect of all this, as can be imagined, was one of unrelieved distaste. The popular view that public schools are luxurious resorts for the pampered children of the affluent classes is entirely misconceived. Indeed, at least when I was there, the whole atmosphere was spartan and strict. Though I carefully refrained from giving any hint of this in my letters to my parents (files of the ones I wrote to my father came to me after his death), I vividly remember counting the days for term to end from the very first day, and also the feeling of hopeless resignation in the pit of my stomach every time I left home after the holidays.

The school routine was full, both with academic work as well as sports and hobbies. Although there was nothing extraordinary about the routine, I will describe it briefly because it contrasted sharply with my previous existence and involved considerable adjustment on my part. We had to get up at six to the sound of the huge gong rung by the resident chowkidar. Chota Hazri—a cup of milk and a slice of bread—would be followed by P.T. on the main field. While I somehow managed most of the drill, I was particularly mortified during the weekly visit to the gymnasium where, for some reason, I developed an irrational fear of doing a forward roll on the box. Whenever I mounted the box for the purpose I froze, and nothing would persuade me to do the somersault. Time and again I would find myself in the absurd situation of sitting on that confounded box with the whole class sniggering at my embarrassment. Even a talking to by the Headmaster, the formidable Mr A. E. Foot, did not have any effect, and throughout my school career I never managed to break that particular hoodoo. After P.T., which the whole school did together, the boys returned to their respective houses. There were four—Kashmir, Hyderabad, Jaipur and Tata—and it was in our houses that we spent most of the time. A wash, change, breakfast and then we were off to the school building for Assembly and classes where each teacher had an allotted room. After the morning classes we would return to our houses for lunch, some rest and then attend the afternoon classes. After tea there would be sports, in which every boy had to participate, then back to the houses for a bath, evening homework (for some strange reason known as 'toye-time'), dinner and then bed.

The academic standards at school were high, but I was always a reasonably quick learner and able to cope quite satisfactorily despite my age. There were as many as nine subjects for the Senior Cambridge, the examination for which the school prepared us. Teaching standards were generally satisfactory, but there were few teachers who left a permament impression on me. The most memorable was Mr V. S. Chari (later V. Siddarthachari of the Indian Foreign Service), who took us for English—my best subject—and was excellent. It was he who first introduced me to English poetry, and even if the class bell sounded to mark the end of his period, none of us moved a muscle until he had finished. The boys were not nearly as well-behaved with all the teachers, and this taught me the useful lesson that only by maintaining one's competence and dignity could one elicit the respect of others; being too lenient and indecisive was invariably exploited, specially by the bullies and roughnecks.

True to the public school obsession with physical fitness, there was a good deal of stress on excelling at games, team events as well as individual. The four houses competed fiercely against one another in every game, and the whole school took part in the annual Dehra Dun district sports meet. I was never in any of the school teams except chess, which I captained for two years. The sport I liked least was cross-country running, a dreadful ordeal in which we had to cover miles over the hillocks and ravines that surrounded the school, sides aching and lungs bursting. The school also provided a wide variety of hobbies from which we had to choose two. I chose music and carpentry, the latter because everyone else seemed to be doing it. Each boy had a Report Card which had to be filled in by his teachers every month. The system of punishments was interesting and unusual. Corporal punishment was forbidden, and the severest punishment (short of expulsion from school for really serious offences) was a Yellow Card. This involved an announcement on the school notice board, forgoing Tuck Shop (a much treasured privilege) for a month, no cinema (to which we were taken occasionally) for the next two occasions, and so on. A Red Card was for consistently bad performance in studies and also carried a set of liabilities, and a Blue Card could be given by house prefects for minor misdemeanours. I went through my entire school career with only one Yellow Card, which I got because the boy sitting next to me during a history test copied my faulty answer, as a result of which we both got caught.

In the middle of every term there was a three-day break during which we all went on various expeditions to the many attractive locations in which the Doon valley abounds—Doiwala, Lachiwala, Raiwala. The more adventurous older boys went off on climbing expeditions, but my own ventures were confined to rafting down the rivers nearby or simply hanging around the picnic site. I was a non-swimmer when I joined school, and it was some years before I overcame the fear of water and diving. In a public school timidity is not a particularly prized asset, and looking back it is clear to me that this deprived me of enjoying school as much as I might have. Unfortunately the original Kashmir housemaster, Jack Gibson, left for naval service the very term I joined, and returned only in my last term. The Headmaster, Mr Foot, was aloof and forbidding and in any case the school was structured in such a way as to make the house the real focus of a boy's life. In the result, while the age factor denied me the satisfaction of a large circle of friends, I also did not receive the benefit of contact with an outstanding housemaster. Added to this was the fact that my father had forbidden anyone to visit me while at school, and that there was no one with whom I could discuss my problems and uncertainties.

School did, however, have a silver lining in the shape of bi-annual vacations—two and a half months in summer and six weeks in winter. After I joined Doon my separate establishment at home was wound up, Digby and Bilti were sent back to their homes, and I began spending the holidays with my parents. These years, from 1943 to 1946, were the closest I got to a normal family atmosphere in childhood, as they also happened to coincide with a phase of comparative harmony between my parents. The summer vacations were spent in Srinagar, where we all lived in the main palace known as Gulab Bhavan, now the Oberoi Palace Hotel. This handsome double-storey structure built on three sides of a rectangle commands a superb view of the Dal Lake and reflected my father's keen architectural interests. Unlike many contemporary princes, whose palaces were vast Victorian monstrosities, our residence with its clean lines and uncluttered exterior fitted in well with the mountains that formed its backdrop. The forests came right down to the back of the building; my father once shot a panther from the main lawn, while the late Maharaja Sardul Singh of Bikaner shot a bear from his bathroom window. ('One bare body against another', as a friend remarked in later years.)

The front lawns were carefully manicured, and beds of multi-coloured flowers shone brightly in the crisp Kashmir air, distinctly crisper than it now is, many decades later. The palace was tastefully furnished, with exquisite European and Chinese curios, and wall-to-wall carpeting in every room and corridor. A lot of furniture came periodically from England through a Mrs Sutherland, the widow of an Englishman who had been Chief of the Army Staff in the State at the crucial juncture around the time of my birth, when the Muslim Conference under Sheikh Abdullah first launched its anti-Dogra movement. Apart from the carpets, which were made in Kashmir, all the other furnishings, fabrics and fittings were European. My parents occupied inter-connecting wings on the first floor of the southern block, while I had a lovely suite on the ground floor directly overlooking the lake. It was only after being exposed to the grim surroundings of the Doon School that I began to appreciate the quiet elegance and beauty of our houses which I had thus far taken for granted.

It now became evident that my father also looked forward to my vacations. We would go out on fishing trips and to Dachigam for shooting-cum-picnics, and played lots of rummy, backgammon and ludo sitting out in the front lawn under one of the Chinar trees. Victor Rosenthal and the Stilemans were usually up in Srinagar at the time, which added to our fun. We spent hours working out lists for lunch and dinner parties, seating plans and menus. Both my parents were excellent cooks, and at least once a week there would be a cooking party. We all sat out in the lawn or the black marble porch in the centre of the main block, and then the cooking would begin. Liveried servants with yellow turbans brought in low asbestos stands about six inches high and five by three feet in area. On these would be placed coal stoves, and on them silver *patilas*, deep cooking dishes of various sizes depending on the item to be prepared. All the ingredients, carefully weighed and neatly arranged, were brought on folding tables. Three or four people cooked, including my parents, while the others watched and chatted. The chief taster was Wazir Tej Ram, a seasoned courtier a few years older than my father who, along with Faqir Singh, was a regular dinner guest at the palace.

Whisky would be served in small glasses, but I seldom saw my father taking more than his standard quota of three double pegs. This was when he would loosen up a bit and relax, and unless some staff problem erupted the atmosphere was comparatively free of tension. The cooking-cum-drinking-cum-gossip parties would last for several hours, dinner not being served till around eleven. I would sit in for the first couple of hours, then my dinner was brought on a tray at nine and I would eat while the rest kept up the party. After dinner I was allowed to stay on for a few minutes, and then at nine-thirty I withdrew, taking leave of my parents by touching their feet and bowing to the rest of the company. I would go off to my room and fall contentedly asleep with the sound of the party wafting faintly across the lawn.

I did not have any friends of my own age, but cousins from my mother's side were allowed to come and play with me from time to time. I met the girls at my mother's apartments, while the boys came to my room or we would play in the garden. The oldest boy, Nasib Chand, a few years older than I, was my favourite and the closest I had to a real friend.

During the summer holidays there were occasional official functions, which my father detested. He was a strange person;

highly intelligent and talented in many ways, but shy and uneasy in public. In fact he often said to his friends that he was only waiting for me to be twenty-one so that he could hand over the State responsibilities and then do the things he loved—shoot, fish, cook, race and build. It is a strange irony that twenty-one was in fact the age at which I took over from him what remained of his authority, but under circumstances that could not have been even remotely imagined at the time.

My father started training me early for public life. My first public speech was at the age of eleven, when I opened the annual exhibition at Srinagar, a combination of a trade and industries fair, entertainment complex and general fiesta, which had become an annual event. The opening took place during my summer vacation, so it became a regular fixture for me to perform the inauguration. I recall vividly reading out my first speech, heart quaking inwardly but managing somehow to display a blasé exterior. After the speech Colonel Haksar took me to Dachigam where my father awaited me with obvious pride. He was not a demonstrative person, but it was not difficult to gauge his feelings at any given moment. Some years later, in 1945, Gopalaswami Ayyangar recalled this incident when he wrote in a letter to my mother: 'Tiger must be quite a young man now. . . . I well remember the first public speech he made when I was in the State and how greatly it was appreciated for the correct enunciation and self-confidence with which he delivered it to a large audience.'

While the summer vacations were spent in Srinagar, the winter holidays were in Bombay, where my parents would go for several months of the year. In Bombay our whole life revolved around racing. My father was for years a leading patron of the turf, and maintained a string of superb race horses. These were stabled in the Mahalakshmi area near the racecourse, and every evening my father went to the stables and watched his horses go round and round the enclosure. There were gallops and spurts, small race books and large ones, confabulations with jockeys and trainers, thinly veiled animosity towards rival owners and, finally, race day on Saturday or Sunday. It seemed as if the whole week was a preparation for this event. My father invariably had at least two horses running, often as many as four. His scarlet and gold colours were familiar to every

Bombay race-goer, and for many years he was the leading owner, bagging many of the most prestigious races. He was a fine judge of horse flesh, and took great pains to select youngsters at the annual bloodstock sales in Bombay. As a logical culmination of this he started a stud farm of his own at Nagbani, a few miles outside Jammu, where by the time he died he had assembled some of the best stallions and brood mares in the country. Quite clearly, my father was much happier racing than administering the State, which chore he left largely to his carefully chosen Prime Minister and a small Council of Ministers, mostly from outside Jammu and Kashmir. In fact, although he enjoyed absolute power, he conducted himself rather like a constitutional monarch and hardly ever interfered with the administration of his Council of Ministers. In this regard he was far in advance of most of his princely contemporaries in India.

For me the holidays were an eagerly awaited event. They flashed by much too fast, unlike terms at school which seemed unending. Quite apart from superspace and time-warps, I learnt early that time does not flow at a uniform rate, at least for a homesick schoolboy. In due course I gradually began to get more adjusted to school life, though I did not make many abiding friendships. In fact, I generally disliked being at school, but I can see quite clearly that, had I remained at home or gone to one of the feudal establishments, I would probably have been hopelessly spoilt and quite unfit to meet the challenges that—unknown to me—lay only a few years ahead.

My only regret is that because of my father's strictness and the absence of an outstanding housemaster, I did not have any strong figure upon whom I could rely during those formative years, someone who might have filled the inner vacuum of uneasy insecurity that afflicted me. I remember once how a casual remark by Pandit Ramchandra Kak (appointed my guardian after Colonel Haksar) to the effect that he had great hopes and faith in my future as Yuvaraj gave me tremendous sustenance for months. Pandit Kak was in many ways a remarkable man, arrogant but unflinching in his adherence to the principles in which he believed. He would tell me that the greatest quality one could develop was 'poise', a calm imperturbability in the face of any circumstance, howsoever unsettling, and he displayed this himself in the adversity that he faced only a few years later.

Meanwhile, I had entered my teens, and the eternal mystery of the human body had begun to unfold itself as it has done for aeons, the same yet ever new. Those are breathtaking years for a boy, the first, tentative probings into manhood, the stunning discovery of hitherto unimagined dimensions of the human body. Kashmir is a marvellous place for such an awakening; the air crisp and cool, the doves cooing melodiously in the distance, the great Chinars rearing their heads high against a powder blue sky, and a boy on the threshold of manhood—bewildered, awestruck, ecstatic.

I had developed the habit of reading when I was quite small, and by the time I was finishing school I was a prolific reader. My favourites at that time were the thrilling Scarlet Pimpernel novels of Baroness Orczy and the hilarious Jeeves stories of P. G. Wodehouse. This apart, there were Dickens, Scott, Hardy and other British classics; Maupassant, Dumas and Victor Hugo among the French; Tolstoy, Chekhov and Gorki among the Russians. Poetry I always loved, and at one stage knew dozens of poems from *Palgrave's Golden Treasury* by heart. I discovered soon that I had a sharper memory than most other boys, specially when it came to verse. I think this was due to the rhythmic aspects of poetry, so closely allied to music. I was also interested in acting, and took a fairly active part in house and school plays. My high point was portraying Olivia in *Twelfth Night*, happily well before my voice cracked, and later I began directing plays with smaller boys in them.

Films were also an important element in my life at that stage. Apart from oriental fantasies with John Hal, Maria Montez and Sabu, there were remarkable Indian films like *Bharat Milap* and *Ram Rajya*, Vijay Bhat's famous historical films which brought the Ramayana to life for a whole generation; *Sikander*, with Sohrab Modi and Prithviraj; *Pukar*, with Naseem and Sohrab Modi; *Shakuntala* with Jayshree and Chandramohan; *Rajputani* with Bipin Gupta's splendid portrayal of Rana Pratap; *Chandragupta* with Nayampalli as an unforgettable Chanakya, and many others. The abiding impact of films on impressionable young minds is often underestimated by adults, with malign results that now speak for themselves.

Despite a few bad patches I was able to do well in my studies, and as the major examination neared I got better and more self-confident. I sat for the Senior Cambridge examination in December 1945, the year the Second World War ended. The centre was in our own school, but there were invigilators from outside. The papers in those days were set in England and corrected there, with the exception of Hindi. By this time I had outgrown many of my early inhibitions, and was for the first time actually beginning to enjoy being at school. There was a further examination offered by the Doon School at that time, the Intermediate Examination of the U.P. Board of Education. This was divided into Inter-Science and Inter-Arts, and almost invariably the brighter boys chose Science. I was quite clear, however, that for the public life to which I appeared destined Arts, which included economics, civics and history, would be more suitable. When I returned for the first term of 1946 after the Senior Cambridge, therefore, I joined Inter-Arts and found myself at one stroke head and shoulders above the rest of the class, all the brighter boys having predictably chosen Science. The Senior Cambridge results were announced during the term, and I secured a first division. I sent a telegram—my first—home, and was soon inundated by congratulatory telegrams not only from my father but from many of his loyal subjects. The school office had never before received so many telegrams, and this caused quite a flutter. I must admit that I was rather pleased with myself, particularly as I had been able to overcome the two-year age barrier and do better than many older boys in my class.

It was a measure of the isolation of our school from the winds of change blowing across the subcontinent that almost till our last year we were barely aware of the political developments that were gathering momentum and would within a year create an India free but divided. Although we read first about the Cripps Mission and then the Cabinet Mission led by a worthy gentleman whom we revelled in calling Lord 'Pathetic' Lawrence, we did not have any real awareness of the tremendous forces that were at work to create a new era in modern history. Our sympathies were, naturally, with Gandhiji, but the boys at school were mainly children of Indian Civil Servants, Service Officers and rich businessmen, who could hardly be expected to be in the mainstream of the national movement.

At just about this time, I came across Jawaharlal Nehru's *Autobiography*. I was enthralled. Here was an intelligent and sensitive man, born to comfort but aligning himself fully with the hopes and aspirations of millions. Reading that book at that particular moment was a revelation. It brought home to me for the first time some awareness of the power of historical forces, of the inevitability of change, of the grandeur of a national liberation movement. I followed this up rapidly with his *Discovery of India*, which opened up for me a new world of the mind. I had, of course, a generalized pride in being Indian, but never before had the long and rich panorama of our history unrolled before me, nor had I appreciated the rich diversity and almost miraculous unity that had characterized India from the very dawn of human history. Reading Jawaharlal's two books gave me a new awareness, and strengthened the aversion to feudalism that had already been growing within me. I realized clearly that the old feudal structure was on the verge of collapse, and that, in any case, my father's life was not for me. What the alternative was I did not know, but inwardly I was ready for change.

Chapter 4

The storm in fact was almost ready to break. Centuries of Western colonialism were on the verge of ending, and the movement towards freedom of hitherto subject peoples—the greatest of its kind in the long panorama of human history—was nearing a triumphant conclusion with India in the vanguard. Something deep was stirring in our vast and ancient land. India was rising again, the miracle of renewal was once again ready to occur. The most destructive war ever fought by mankind had just ended. My father had gone to England in the course of the war as a member of Winston Churchill's War Cabinet, evidently one of those elaborate farces at which the imperial British excelled, aimed at giving members of the British Empire some sort of symbolic say in the affairs of State without any real power to influence decisions. He was in London for some of the worst German bombing, and on his return related to us how, despite the air raid sirens, not a single person would leave his theatre seat for the shelters. Earlier he had paid a visit to the State Forces in the Middle East, and on his return had been given a rousing reception by the people.

My mother, also, was very active during the war days. She organized a War Aid Committee and she and the leading ladies of Srinagar would meet regularly to knit, sew and prepare pickles for troops in the forward areas. Indeed she did so well that she was awarded 'The Crown of India', a decoration reserved for distinguished women and received earlier only by two or three other Indians. She was, of course, thrilled with the whole affair, specially so as in the same year's imperial honours list my father received a decoration that was slightly lower in the hierarchy of protocol. She was also very advanced in some of her views regarding social customs. In 1947 she created history by including Harijan girls, generally regarded at that time as untouchable, among the nine who are worshipped during the sacred *Navratras*.

My first term in the Intermediate class, which I enjoyed more than any earlier one, was marred only by an unexpected development that was to have a great effect on my life. I gradually began developing a sharp, biting pain in my right hip whenever my leg would fall outwards in sleep. I never discovered what caused this, perhaps some trauma during all those riding falls I had earlier, but as the frequency of the pain increased I began to feel more and more alarmed. I did not tell anyone about it for months, a mistake understandable enough for a boy of fifteen, but one that cost me a stiff hip for the rest of my life.

The term ended and we left for our respective homes, fully expecting to return when the next term was due. However, by mid-1946 the communal situation in India had greatly deteriorated. Hindu–Muslim riots broke out with increasing frequency and ferocity, and with the stage set for the British withdrawal the Congress and the Muslim League were locked in deadly combat. Some vague threats about the possibility of my being kidnapped or harmed in some way at school had begun to circulate during that last term, but, of course, we pooh-poohed the whole matter. None the less, my father—aided and abetted no doubt by my mother—decided that it was too risky to send their only son and heir to a far-away school when the country was on the verge of a major cataclysm. Perhaps they were right, but when I learnt of the decision soon after I came to Srinagar for the summer vacation I was very upset. It was, I thought, strange irony that in the years when I detested going to school I was ruthlessly packed off to it, and just when I was beginning to enjoy it was peremptorily withdrawn.

However, as usual, I had little say in the matter, and when my mother ecstatically broke the news to me I didn't have the heart to make a scene. Thus ended my four years at the Doon School, years of painful adjustment and unappetizing food, but also of valuable training in self-reliance which stood me in good stead for the rest of my life. I am not overly sentimental about the old school tie, but I do acknowledge a debt of gratitude to my father for having sent me to the Doon, and to the school itself for having, in its own rather rough way, saved me from growing up in a feudal environment and given me a useful intellectual grounding.

Although my stay at the Doon had been disrupted, I was keen to continue my studies and persuaded my father to send me to the local Sri Pratap College in Srinagar. This he agreed to do, an event which at the time was regarded as a progressive and democratic gesture. I went to college every day accompanied by an ADC, but sat in the classroom with the other students. I took English, civics, history and economics, and also participated in debates and declamation contests. On one occasion I won a declamation prize, which was presented by my mother at the annual prize-giving. Unfortunately, my stay at college turned out to be very short, confined only to the summer of 1946. My tutor at that time was Professor B. K. Madan, a genial and shrewd Kashmiri Pandit who, though no genius, was generally alert and well informed. His main lesson to me, quite in keeping with his character, was that packaging was more important than content. 'Tissue paper and tin foil', he would say, 'is all that really matters; it is *how* a thing is presented that really impresses most people, only few care to find out *what* in fact it really contains.' Cynical doctrine, no doubt, but not devoid of usefulness in this imperfect world.

Meanwhile, a strange development took place in our household. A certain Swami Sant Dev, who had lived in the State decades earlier in the time of the late ruler, Maharaja Pratap Singh, and was reported to have been banished by my father when he ascended the throne, staged a mysterious comeback. He came on to the scene around 1944, and by 1946 had firmly ensconced himself as Rajguru and was installed by my father in the beautiful Chashmashahi Guest House in Srinagar and the house where I lived as a child in Jammu. My father was far from being a religious man, but to everyone's amazement he suddenly became a devout disciple of Swamiji, sitting for long periods on the ground before him and never smoking in his presence. Swamiji was presented by him with lovely silk robes, a silver hookah and many other amenities, including a car. He was a remarkable man in many ways, erudite in several fields of knowledge and pink-complexioned even in his advanced age. He would never actually reveal how old he was, but it was rumoured that he was well over eighty (some claimed a hundred). He took opium regularly and would often nod off to sleep, a phenomenon interpreted by his followers as proof of his direct contact with God.

Much maligned though he subsequently was, I feel that his influence on my father's personal life was all to the good. He urged him to cut down on his smoking and drinking, and generally tempered his rather aggressive nature with some sort of religious commitment, howsoever superficial. Also, he was an influence in bringing my parents closer to each other. My mother being deeply religious was, of course, delighted at this turn of events, and her elder brother, Thakur Nachint Chand—who was for many years the Deorhi Officer or Chamberlain—grew in importance at the court as an intermediary between Swamiji and my father. I also had to see Swamiji regularly, but what with my hip pain was beginning to find it increasingly difficult to sit crosslegged on the floor, and so rather avoided going there. Swamiji always treated me with great affection. He was a connoisseur of classical music, and enjoyed hearing me sing his favourite raga, Jaijaiwanti.

It was in the political sphere that Swamiji's influence proved to be disastrous. As with many of the larger Indian States, the prospect of becoming an independent ruler after the British withdrawal was an alluring one for my father. He had never taken happily to British suzerainty, and at the same time was far too deeply enmeshed in the feudal tradition to really come to terms with the democratic forces that were gaining momentum in the subcontinent and within the State itself. A major weakness of the feudal system is that the rulers are told only what their courtiers think they want to hear, which is seldom in consonance with what is in fact happening. It was on this feudal ambition that Swamiji astutely played, planting in my father's mind visions of an extended kingdom sweeping down to Lahore itself, where our ancestor Maharaja Gulab Singh and his brothers Raja Dhyan Singh and Raja Suchet Singh had played such a crucial role a century earlier. There is also some reason to believe that Swamiji was in touch with some of the Congress leaders, and that Acharya Kripalani's visit to the State early in 1947 was a direct result of his intervention.

Be that as it may, the end of Dogra rule was fast approaching, although my father seemed to be blissfully unaware of the tremendous forces that were on the move in the subcontinent. Within the State Sheikh Abdullah had converted his Muslim Conference, founded in 1931, to the National Conference and had developed personal and ideological links with Jawaharlal Nehru. He was active in the State's people's movement which was a sort of counterpart in the Indian States of the wider anti-British movement of the Indian National Congress. Although Abdullah was never really trusted by the more conservative Congress leaders like Sardar Patel, he succeeded in gaining the close confidence of Jawaharlal who, perhaps because of his Kashmiri origin, looked upon Kashmir as his special sphere of interest.

Jinnah had little use for Abdullah and his associates, partly because the Sheikh was not prepared to toe his line and partly because of Jinnah's messianic belief that he and his Muslim League were the sole guardians of Muslim interests in the subcontinent. Abdullah's tactics were simple but effective. Surcharged as he always was with anti-Dogra feelings, he played skilfully on Nehru's anti-feudalism by making my father the main target of his political assault. In May 1946 Sheikh Abdullah and his National Conference launched a 'Quit Kashmir' agitation on the lines of Gandhiji's celebrated 'Quit India' movement four years earlier. Prior to that my father had taken some steps towards popular government in the State, including a dyarchy experiment in 1944 with the setting up of a partially elected Praja Sabha. But this did not satisfy the National Conference's demand for a full share in State power. In 1945 the Sopore session of that party had been attended by Jawaharlal Nehru and Khan Abdul Ghaffar Khan, and the Indian National Congress, which was generally backing the State People's Conference, had shown special interest in Kashmir in view of the impending partition and the close personal friendship between Jawaharlal Nehru and Sheikh Abdullah. Unfortunately, my father was not able to grasp the historic dimensions of the changes that were around the corner. This is amply borne out by the fact that when on 20 May Sheikh Abdullah and a number of other National Conference workers were arrested, and Jawaharlal Nehru announced his intention to visit the State, his entry was banned. Characteristically, Jawaharlal decided to defy the ban and enter Kashmir from the Panjab via the Kohala bridge. A company of State Forces troops commanded by a Gurkha officer, Major Bhagwan Singh, stood with bayonets fixed half-way across the bridge. Jawaharlal, fearless as ever, walked across and pushed the bayonets aside. By a stroke of good fortune Bhagwan Singh was a tactful and intelligent man. He ordered his troops aside, politely asked Jawaharlal to come across the bridge and informed him that he was under arrest.

The news reached us in Gulab Bhavan, and Pandit Kak, the then Prime Minister, reported to my father with subdued pride and excitement that Jawaharlal had been arrested. I was stunned. Here was the most charismatic leader of the national movement for freedom, the author of the *Autobiography* and *Discovery of India*, the declared future Prime Minister of the Indian Republic, and, instead of welcoming him and seeking his co-operation, we had arrested him. I have no doubt that his arrest was the turning point in the history of the State. Jawaharlal was detained in Uri, and after three days the Congress Working Committee persuaded him to come back to India. Later in July the State Government lifted the ban on his entry and he did come and interview Sheikh Abdullah in jail. But it was too late; the die had been cast, and it was only a matter of time before disaster struck.

It has always seemed to me tragic that a man as intelligent as my father, and in many ways as constitutional and progressive, should have in those last years so grievously misjudged the political situation in the country. He was generally an enlightened ruler, having, for example, thrown all temples in the State open to Harijans as far back as 1932. On that occasion the Raj Pandit in Jammu, who was head priest of our family temple the Sri Raghunath Mandir, had opposed the move. My father dismissed him peremptorily, and appointed his brother in his place only after he had unequivocally agreed to accept the Harijans. Similarly, in many matters relating to land tenure and administration my father had introduced reforms far in advance of most other Indian States. Having appointed his Ministers, the Muslim contingent usually from U.P., he would not interfere with their working. Indeed, there was a case when a citizen of Jammu living just outside our palace challenged in court my father's decision to acquire his property, and successfully prevented the move. In fact when at the age of thirty he ascended the throne on 23 September 1925 he had proclaimed 'Justice is my religion and merit alone will be considered the deciding factor for all types of employment. Caste, creed, religion or sex will receive no consideration.' During his reign he introduced a number of administrative and political reforms, culminating in 1944 in an experiment in dyarchy. Under this the Praja Sabha, set up a decade earlier, was given the right to nominate a panel of six, out of which he would choose two persons to join the Council of Ministers. This, for the first time, introduced a 'popular' element in the government, and towards the end of 1944 two such Ministers were appointed, Mirza Afjal Beg and Wazir Ganga Ram, who had secured the highest votes among the Muslim and Hindu candidates respectively.

Again, unlike most other rulers, my father made a clear distinction between his private property, including jewellery, and State property. He left family jewellery, shawls, carpets and regalia worth crores with the State *Toshakhana* (Treasury), which most others in his place would have appropriated without turning a hair. He never harassed the population for personal ends, living an aloof and self-contained life in the palace. His administration and system of justice is to this day accepted by impartial observers as having been much fairer than those of the post-1947 period. Corruption was far less, and severely punished whenever it came to light.

Being a progressive ruler was one thing; coping with a once-in-a-millennium historical phenomenon was another. There were four major forces at work on the subcontinent at that time, and my father was on hostile terms with each one of them. There were the British, ready at last to surrender the brightest jewel in their Empire. Although he never really believed until the very end that they would actually go, my father was too much of a patriot to strike any sort of surreptitious deal with them. There was the Indian National Congress, inspired by Gandhiji and led by Jawaharlal, Vallabhbhai Patel, Maulana Azad and other giants of the freedom movement. To this party my father was hostile mainly because of Jawaharlal's close association with his arch enemy Sheikh Abdullah. Then there was the Muslim League led by Mohammad Ali Jinnah. Although this party supported the right of rulers to decide the fate of their respective States, and in Kashmir opposed Sheikh Abdullah and his National Conference tooth and nail, my father was enough of a Hindu not to be able to stomach the aggressive Muslim communalism of the League, and this led him to spurn tempting offers from Pakistan. Finally, there was the main political party in the State itself, the National Conference led by Sheikh Abdullah, with whom my father was at loggerheads for decades because he looked upon it as the major threat to his throne and Dogra rule. The net result was that, when the crucial moment came, all the forces that counted were arrayed on the opposite side of the fence. Added to this was my father's disinclination to take a firm decision one way or the other. Thus he found himself alone and friendless, and the edifice of Dogra power built up so painstakingly over a century collapsed. If only I had been ten years older, I felt, I could have changed history. But if I *had* been ten years older, might I also not have fallen victim to the feudal virus?

Despite the excitement caused by the arrest of Nehru, things seemed to settle down again for a while. I spent that summer going to college in Srinagar and, on weekends, accompanying my parents on shooting and fishing expeditions. I was allowed for the first time to try my hand at big game. My father gave me a .318 rifle, and for my first major shoot I was sent to Dachigam with Thakur Harnam Singh Pathania, a crack shot who was a senior officer in our forest service and later rose to be Chief Conservator. We set out early in the morning, and entered the reserve before sunrise. The cold air was fragrant with the aroma of the forest, and the mountains seemed alive and vibrant in that early dawn. I can still savour the sense of exhilaration that I felt as, gun in hand, we made our way through the forest. The sun came up after a while, setting the forests on the surrounding mountain slopes ablaze in their autumn colours. Our quarry was the magnificent Kashmir stag, but we did not come across one. A few doe, their eyes gentle and liquid, stood frozen until we approached quite near and then bolted off into the forest.

At around ten, after we had driven in the reserve a good deal, Harnam Singh suddenly caught my hand and whispered, 'Would you like to have a shot at a bhaloo?' About a hundred yards away, partially obscured by some trees, stood a huge black bear. I raised my rifle and fired. The bullet knocked the bear clean over, but they are tough creatures and, unless mortally wounded, are difficult to recover. Cautiously we moved towards the spot where the bear had fallen, Harnam Singh keeping his rifle at the ready. We found blood but no trace of the bear. It is said that bears have secret hideouts where they go when ill or wounded, and that they can treat their wounds with healing herbs from the forests. Be that as it may, we never found the bear. Perhaps it was symbolic; I was not destined to be a hunter, and some years later I gave up shooting and fishing for good.

The summer rounded off into the superb Kashmir autumn; the Chinar leaves began to turn russet, the paddy ripened and its golden fields shone against the green of the valley. The air was crisp with expectation. Great things were happening on the subcontinent; the Cabinet mission came and went, Jinnah's demand for a divided India shook the subcontinent to its foundations. Lord Wavell, the square, one-eyed Viceroy was replaced by the dashing and handsome Lord Mountbatten and his beautiful wife Edwina. History was in the wings waiting to make a dramatic appearance. And yet we were almost entirely bogged down in petty affairs—the ridiculous little intrigues at my father's court, the circle of small-minded men surrounding him, the coterie of time-servers and yes-men hanging around interminably.

Swamiji's influence continued to grow, and with it the rise to power in court of my mother's elder brother Thakur Nachint Chand. He had been a non-commissioned officer in the Dogra regiment when my mother's marriage took place and, as was customary, had been given a large jagir and made a ranking courtier after the event. He was for many years the Chamberlain to my mother, over whom he wielded great influence, being many years older and having once, when she was a child, saved her from drowning in the village pool at Bijeypur. While he was a shrewd and basically loyal person, his intellectual horizons were limited and he was greatly given to seeking the magic favours of Swamis and Sadhus. There was no doubt that in the very difficult years that my mother spent after her marriage, a village girl suddenly transported into the vortex of court intrigue, he was a tower of strength to her, and he seldom lost an opportunity of pointing out how, had it not been for him, she would have fallen victim and perished long before I was born. In this he incurred the displeasure of the opposing faction at court, and was generally unpopular and disliked by the Jamwal baradari, my father's kinsmen. With the advent of Swamiji, however, he suddenly rose to prominence and became the main courier between him and my father.

We moved down to Jammu for the season as usual in the first week of November. My tutor Professor Madan had made inquiries and found that I could take the Intermediate examination of Allahabad University as a private student. My father was not very keen on the idea, he himself never having gone beyond Matriculation at Mayo College, but Madan and my mother succeeded in persuading him to agree. For this I owe a deep debt of gratitude to Madan Sahib, for if my studies had broken off at that point it would have been impossible for me to take them up again as I did later. A special examination centre was opened in one of our Jammu staff houses, and early in 1947 I appeared for the examination taking English, civics, economics and Hindi. Considering the fact that after the Senior Cambridge my studies had consisted of a term at the Doon, a term at Srinagar College and a few months of rather hurried private coaching at Jammu, I was quite pleased to get through the examination in the second division.

I was growing up rapidly and, looking back, I can see that the corrosive influence of a feudal court had begun insidiously to influence my behaviour. I began to get overbearing in my attitude to the servants, to treat the dogs roughly and generally to start developing those traits that constitute the less desirable features of feudal life. Being deprived of virtually any company of my age—Nasib was away in college at Lucknow—I spent more and more time with the staff. My mother and father seemed to be coming somewhat closer, though the basic tension remained, often bursting out into searing scenes which left me deeply mortified. My mother at this time lived in Amar Mahal, the large brick chateau that my grandfather had started building in Jammu but never completed, while my father lived in the low, grey house next to it known as Hari Niwas, where I also had a room. My mother would come across for meals, and we would all play a good deal of rummy and backgammon. Our life in Jammu was even more isolated than in Srinagar. The only place we visited was Udhampur, a town forty miles from

Jammu, where my father had developed some fine shooting reserves. He also built a residence there named—to my pleasure and surprise—Tara Niwas, after my mother. There we would all stay before going out to the various *rakhs* for shoots. Though lacking the magnificence of the Kashmir mountains, the hills around Udhampur nurtured a wide variety of large and small game including panther and wild boar.

All these months the pain in my right hip joint had continued to increase steadily. I began to dread falling asleep at night because of the sharp, shooting pain that would occur. The muscles of my right hip and leg began to atrophy, as a result of which I began limping slightly. Whenever my mother remarked on this I would brush the matter aside, but one day my father noticed it also and asked the Palace physician, Dr S. K. Shangloo, to examine me. It was then that he discovered the muscle atrophy. He started me on some oil massage for a few weeks, but it made no difference. Ultimately my father decided that I should go to Bombay for specialist consultation. The Controller Household, Brigadier N. S Rawat, accompanied me and I went through a series of tests and X-rays, including a visit to the famous Dr Khanolkar, founder of the Tata Cancer Institute. Evidently someone had suspected cancer, but this was ruled out. We were advised that if the pain did not disappear within three months it would be necessary to put me into a plaster cast to immobilize the joint for a while.

With this depressing news I returned to Jammu. My father, as usual, spent most of his time from March onwards in the swimming pool where, in the evenings, he played water polo with the staff and selected servants. I was not much of a swimmer, and took only a perfunctory part in the proceedings. I would wander off to Amar Mahal to spend some time with my mother, or down to Karan Niwas where Swamiji lived. In fact, I was at a loose end, and deeply disturbed by the hip pain and the general atmosphere. The summer grew upon us, but my father showed little inclination to fix a date for the move to Srinagar. Ultimately, my mother intervened, saying that it was getting intolerably hot in Jammu. This led to a strangely sharp reaction from my father, who took the suggestion with very ill grace and sulked about it for days. Looking back, I wonder whether he had some sort of premonition that this was to be his last visit to Kashmir as ruler.

We finally got up to Srinagar towards the end of May, and immediately plunged into hectic activity in connection with the forthcoming visit of the new Viceroy, Lord Louis Mountbatten, and Lady Mountbatten. Viceregal visits to Indian States were a regular feature of British rule, and I vaguely remember Lord Linlithgow having come sometime in the thirties when I was a small boy and presenting me with a gold-ringed riding stick. But this visit was special, not only because the British had already announced that they would leave India soon and hand over power but also, at least as far as I was concerned, because of the personalities involved. Coming as they did soon after the rather dour and unprepossessing Lord Wavell, the glamorous Mountbattens were an exciting contrast and I was eager to meet them.

My father was meticulous in making ceremonial arrangements; programmes were neatly printed and bound in golden paper with a maroon bow (gold and maroon being the State colours), and all details including guest lists and menus worked out down to the last detail. He would spend hours over such details, and the whole household was in a permanent state of tension, for if anything went wrong someone's head was sure to roll. As I was doing nothing particular at the time I also took a hand in helping out with staff duties, although my contribution was, to say the least, marginal. The great day finally came, and the Viceroy and Vicereine arrived to the booming of guns (I can't remember how many, but as my father himself ranked a 21-gun salute I presume the Viceroy got 31). I was at the porch with my mother and father to greet them. My first look confirmed and fulfilled all my expectations. They were every inch a charmed couple, and at once—and irrevocably—I associated them with my favourite characters from the Baroness Orczy novels, Sir Percy Blakeney and his beautiful wife Marguerite. They looked as if they had stepped out of the pages of one of these novels; he tall, handsome, dashing; she beautiful, gracious, charming. My boyish imagination, long in search of a break from the rigid atmosphere of our own court, found in the Mountbattens a marvellous window where fact and fantasy mingled.

Despite their aristocratic background, there was nothing

stuffy or formal about them and I was delighted at their informality and sense of humour. My father had introduced me as 'Tiger' and they called me that throughout the visit and, in fact, ever after. There was a whole series of garden parties and banquets and receptions. Two amusing incidents stand out. At the banquet there was a bell under the table, which my father was supposed to press when the party was over and the band was to play 'God save the King'. Mountbatten, sitting next to him, was tall and his knee hit against the bell by mistake in the course of the dinner, at which point the band dutifully struck up the anthem and all of us struggled gamely to our feet roughly half-way through the chicken curry. My father was furious and went red in the face, but luckily there was no one he could blame. Mountbatten laughed uproariously when he discovered what had happened, and a fit of giggles had me almost rolling on the floor despite my mother's glares from across the table. The second incident was when my father had to introduce his senior officials to the Viceroy before a garden party. He had everyone lined up and had carefully memorized the list of names. For some reason, however, one of the officials stood in the wrong place as a result of which the whole order went awry, so that when my father rattled off the introductions he suddenly realized that he was using the wrong names or, rather, introducing everybody by a false name. I can never decide whether he or the officials were more disconcerted, but Mountbatten didn't seem to notice anything was amiss or, if he did, was far too polite to give any indication to that effect.

Apart from the fun and the festivities, however, Mountbatten's visit had a serious political purpose. The day for the British to depart was rapidly nearing, and while most of the Indian States—by sheer compulsion of geography if nothing else—were making up their minds which of the two new nations to join, there were a few who had still to make their decision. These included the two biggest Indian States, Hyderabad and Kashmir. While Hyderabad was entirely surrounded by territory belonging to the Indian Union, Jammu and Kashmir had boundaries both with the Indian Union and the proposed new State of Pakistan. Further, our State was ethnically diverse and inhabited by Muslims (both Sunni and Shia), Hindus, Buddhists and other religious groups. I suspect that in his heart my father still did not believe that the British would actually leave. Indecisive by nature, he merely played for time.

To be fair to him, of course, the situation he faced was a complex one, and there was no easy option. If he acceded to Pakistan, a large chunk of his people, including his entire Dogra base, would have been outraged. If he acceded to India he risked alienating a large section of his Muslim subjects. Independence could perhaps have been an attractive proposition, but to carry that off would have required careful preparation and prolonged negotiations with the parties concerned, as well as tremendous political and diplomatic ability. Mountbatten in fact had come to persuade my father to make up his mind well before 15 August, and had brought an assurance from the Indian leaders that they would not take objection to his deciding in whatever way he thought fit, even if it was accession to Pakistan.

A typical feudal reaction to a difficult situation is to avoid facing it, and my father was particularly prone to resort to this. Instead of taking advantage of Mountbatten's visit to discuss the whole situation meaningfully and trying to arrive at a rational decision, he first sent the Viceroy out on a prolonged fishing trip to Thricker (where Mountbatten shocked our staff by sunbathing in the nude) and then—having fixed a meeting just before his departure—got out of it on the plea that he had suddenly developed a severe attack of colic. Mountbatten, as has been recorded by his aide Campbell-Johnson, saw through the deception without any difficulty and returned to Delhi. Thus the last real chance of working out a viable political settlement was lost.

Chapter 5

Meanwhile my hip showed no improvement, and ultimately the eminent surgeon Colonel Mirajkar was called in from Lahore. He advised that the hip joint should be immobilized in a plaster cast. The prospect of being confined to bed was highly distasteful, but the pain was getting worse and there was no alternative but to accept the doctor's advice. In June the plaster cast was put on by Mirajkar himself, and I found myself at one stroke converted from a reasonably normal human being into virtually a cripple. For anyone who has not been through the experience, the physical and psychological problems of being confined to bed in plaster for a long period can hardly be imagined. The sheer physical discomfort of the hard cast, which started at my waist and went all the way down the right leg to the toes, was formidable, and the incapacity to move without help even in bed, the itching under the plaster from which there was no possibility of relief, and, above all, the inability to go to the lavatory were a maddening combination.

The nights were the worst. In the day I was lifted out of bed on to a wheel-chair and brought in to sit with my parents and others. They all tried hard to keep me amused, and we would play rummy and ludo. In the evenings some of my cousins would visit me or I would read. But when night fell and I was alone in bed with the dreadful plaster cast, my heart would sink and I had a dreadful feeling of trapped hopelessness. The only consolation came from prayer. My mother had given me a rosary and a picture of Durga—the great goddess riding on a lion—which I kept by my bed. Often I would spend hours telling the beads and sometimes fall asleep with the rosary in my hand. Occasionally I would dream that the whole thing was a nightmare, and that when I awoke I would find myself perfectly fit again. But in the mornings I would awake to the realization that my illness was a grim reality. I think it was in those days that I developed the capacity to come to terms with reality, howsoever unpleasant. I was always introspective, but this enforced and prolonged confinement when boys of my age would be at their most active, drove my mind deeper into itself.

I had been told that the plaster would be on for three months, after which I would be well again and able to resume a normal life. The days lengthened into weeks, the weeks into months. It was 1947, the year in which India was to rise to freedom through an ocean of blood and destruction. Much has been said about the non-violent nature of our freedom struggle, and how we were able to achieve independence without resort to armed conflict. This, of course, is true in the sense that not a drop of British blood was shed in the process, and Gandhiji was certainly a unique leader. But India paid in full measure the bitter price of freedom in the savage and tragic communal conflicts that swept the subcontinent in 1946 and 1947. Hundreds of thousands of innocent men, women and children were sacrificed in the fires of religious hatred and fanaticism as Hindus and Muslims found themselves locked in deadly and unequal struggle—unequal because in all cases it was the minority in any particular city or locality that suffered most. Streams of Hindu and Sikh refugees began to pour into the State from neighbouring Panjab. Jammu and Kashmir had for long been an example of communal harmony, and at a time when the whole of India was aflame it seemed as if the State would remain a haven of peace and tranquillity. This is what Gandhiji meant when he said that in the engulfing darkness the only ray of light seemed to him to come from Kashmir.

Gandhiji in fact visited Srinagar in August that year. There was great excitement when we knew he was coming, and this rose dramatically when we heard that, in disregard of his normal practice, he would come to see my father at the palace. Even my father seemed thrilled, and of course I insisted that I should meet him also. After some discussion it was agreed that my father, my mother and I would meet Gandhiji under one of the Chinar trees in the front lawn of Gulab Bhavan. Special arrangements were made for goat's milk and fruit, and we took our places under the tree an hour before he was due to arrive. At the appointed hour—five in the afternoon of 1 August, if I remember correctly—Gandhiji arrived. My father went to receive him at the porch, and they walked out into the garden to the tree where my mother and I awaited them. I will never forget how moved I was to see that frail figure walking towards us, tiny alongside the tall figure of my father. Here was a man who had become a living legend, who by sheer moral courage had shaken to its foundations the greatest empire the world had ever seen. Though I have a fairly sharp memory I remember hardly anything of the conversation. As he sat down Gandhiji turned to me and said '*Kaise ho*' (how are you?). Then he started a long monologue in a low, lisping voice which I could not clearly follow. My father listened respectfully, but as far as I can recall said hardly anything himself. All I was able to gather from Gandhiji's words was that he was urging my father to ascertain the wishes of his people, to take his people into confidence, and to align himself with rather than against them in the political turmoil that was engulfing the country.

After about ninety minutes Gandhiji got up to go. My mother pressed him to partake of some milk and fruit, but he declined saying that it was not his eating time. At her insistence he agreed to have the fruit put into his car. He left, smiling at me as he said goodbye, dressed in white and with a sad smile which still lingers in my memory. Then he walked back across the lawn followed by both my parents and I was left alone under the tree. I never saw him again in life, but I did see him once in an extraordinarily vivid dream from which his picture still stands in crystal clarity within my mind. But that was several years after he had fallen victim to the assassin's bullets.

Events now began gathering momentum. The British actually seemed to be sincere in their desire to leave by 15 August. It has rightly been said of the British that nothing they did in India became them so much as the manner of their leaving, and it is quite remarkable how all traces of bitterness between us and them disappeared so soon after Independence. This was, of course, largely a tribute to the unique manner in which the Indian National Congress led by Gandhiji conducted the freedom movement. His doctrine of opposition without hatred, and struggle without violence, was matched by the consummate commonsense and the strong liberal tradition of Britain, particularly under the post-War Labour Government. But with the continuing deadlock between the Congress and the Muslim League, the partition of the subcontinent became inevitable.

The position of the five hundred and odd Indian States remained theoretically ambiguous, because the British took the view that the rulers were free to accede to either of the new States that would come into being after they left. However, as Mountbatten pointed out quite bluntly to the Chamber of Princes, their choice was in fact virtually dictated by the compulsions of geography. Whenever these were sought to be violated, as for instance in Junagadh and Hyderabad—both situated entirely within the Indian Union—the result was a foregone conclusion. Not so, however, in the case of Jammu and Kashmir. A bare glance at any map of India at the time is sufficient to highlight the unique geographical location of this State, with boundaries that abutted both India and Pakistan, as also Tibet in the east and separated from the Soviet Union in the north only by a narrow strip of Afghan territory. I have mentioned earlier the special British interest in the northern province of Gilgit, and several historians have built a whole thesis around the consistent determination of the British to dominate this crucially strategic area. Indeed as far back as 1935 my father had been pressurized by the British into leasing the Gilgit Agency to them for sixty years. To add to the complexity of the situation, the State was composed in its various regions of people belonging to vastly different ethnic, cultural and religious groups. Thus the valley was inhabited predominantly by Sunni Muslims with a small community of Shias, Sikhs and the redoubtable Kashmiri Pandits; Jammu predominantly by Dogra Hindus with a significant Muslim component, the western strip from Muzaffarabad to Mirpur by Panjabi Muslims; Gilgit, Skardu and Kargil by Shia Muslims, and Ladakh by Lamaistic Buddhists. This extraordinary patchwork was, of course, the handiwork of my great ancestor Maharaja Gulab Singh who carved out the State with consummate skill in the middle of the nineteenth century. As long as the question of Pakistan was merely an academic exercise, the people of the State remained by and large united in their loyalty to the ruling family, although Sheikh Abdullah's bitter anti-Dogra movement in the thirties and forties had succeeded in making an impression upon politically aware sections in the Kashmir valley. But once the British intention to withdraw became established, and the emergence of Pakistan a certainty, the whole situation underwent a fundamental change. The

State was thrown into inner turmoil; areas bordering the future Pakistani provinces of Panjab and N.W.F.P. started becoming restive, the valley saw the growth of the political movement led by Sheikh Abdullah who was pro-Jawaharlal Nehru but against my father, while the Muslim Conference was pro my father but also pro-Pakistan.

The situation was so complex that even a person far more aware of contemporary realities than my father would have found it virtually impossible to come to a neat, peaceful solution. As I have pointed out, if he had acceded to Pakistan, the Hindu areas of the State would have been virtually liquidated in the wake of the communal frenzy sweeping across north India at the time. If, on the other hand, he had earlier acceded to India he would have run the risk of alienating a large section of his Muslim subjects who constituted seventy-five per cent of the State. In retrospect the only rational solution would appear to have been to take the initiative in promoting and presiding over a peaceful partition of his State between the two new nations. But that would have needed clear political vision and careful planning over many years. As it turned out, the State was, in fact, partitioned, but in a manner that caused untold suffering and bloodshed, poisoning relations between India and Pakistan right down to the present day.

My father's only positive reaction to the rapidly approaching partition was to offer to sign a Standstill Agreement with both the 'Dominions', as they were then called. Pakistan signed the agreement but, soon thereafter, began exerting pressure on the State by interfering with essential supplies. At that time all the major communication systems into the State lay through Pakistan, the road from Kohala and the two railheads at Rawalpindi and Sialkot. Thus Pakistan launched a virtual economic blockade to force my father to accede to it, while India asked for further clarifications before signing the agreement. In a letter replying to a communication from Mehr Chand Mahajan, who had by then taken over as Prime Minister of the State, Jawaharlal Nehru wrote on 20 October 1947:

My dear Mr Mahajan,

I am in receipt of your letter of the 18th October.

I am aware of the difficulties Kashmir has had recently, more especially in regard to the attitude adopted by Pakistan towards it.

When you were here last we discussed this matter also. I assured you then that we have the friendliest feelings towards Kashmir and its people and that we would gladly help to the best of our ability in providing Kashmir with the commodities it specially needs. We would like to do so for humanitarian reasons as well as because of our deep interest in the future of the people of Jammu and Kashmir State. Our self-interest also demands this. We are strongly of the opinion that no coercion should be exercised on Kashmir State and its people and that they should be allowed to function in their own way and to make such decisions as they think fit and proper. In the furtherance of this policy we shall direct our efforts.

You will appreciate that there are some difficulties at present in establishing proper communications between India and Kashmir. We hope that these difficulties will be overcome soon by our joint efforts. In regard to sending necessary commodities to Kashmir, we should like to know what exactly you require. You are aware that the situation in India in regard to commodities is also a serious one and many important articles are rationed because of short supply. It is not easy for us to spare them. Nevertheless, we shall try our utmost to send you anything that you particularly need. I understand that salt and kerosene oil are specially needed by your people. Could you kindly let me have some idea of what you require urgently?

Regarding other forms of assistance, you have our good-will but you will appreciate that these will depend on circumstances. I suggest to you that there should be close contact between Kashmir State authorities and our Government so as to ensure co-operation in matters of common concern.

Yours sincerely,
Jawaharlal Nehru

Intelligence reports from the frontier areas of Poonch and Mirpur as well as the Sialkot sector started coming in which spoke of large-scale massacre, loot and rape of our villagers by aggressive hordes from across the borders. I recall the grim atmosphere that began to engulf us as it gradually became clear that we were losing control of the outer areas. My father occasionally handed some of these reports to me and asked me to explain them in Dogri to my mother, and I still recall my embarrassment in dealing with the word 'rape' for which I could find no acceptable equivalent. Meanwhile all my father's Muslim friends seemed to fall away, including 'Bhaijan' Effendi. My personal grapevine told me that he had, before leaving with his family for Rawalpindi, tried desperately to see my father but had been refused permission to enter the palace gates. I like to think that he was making a last effort to help in averting the disaster that was fast closing in upon us.

A new, unfamiliar set of people began appearing at the palace. Ramchandra Kak, the one man who had the intellectual capacity to make some coherent effort towards an acceptable settlement, had resigned after being humiliated by my father. His place was formally taken by an old Dogra Minister, General Janak Singh Katoch from Kangra, who had served our family with great loyalty for many decades, but it was clear that he was merely a figurehead. Ram Lal Batra appeared from some obscure corner of the Panjab to become Deputy Prime Minister. He had evidently been Dewan in one of the tiny hill States of north India, and his comprehension accurately reflected this background. Yet another person had earlier appeared who was to play a more important role in the events of the next few months—Lala Mehr Chand Mahajan, also from Kangra and who became Prime Minister at this crucial juncture with the blessings of Sardar Patel, although it later became clear that he was not particularly popular with Nehru.

The centre of activity in our house was where my father sat in the lower corridor of the Eastern Wing. After breakfast I was also wheeled in there, my mother came down, the courtiers and advisers assembled and we would sit there throughout the day listening to the radio, sometimes playing ludo, backgammon or rummy. Victor Rosenthal was there, and he did his best to cheer my father up because he had become increasingly withdrawn and hardly ever left the palace. Swami Sant Dev was still ensconced at the Chashmashahi house, but as the situation deteriorated his visits to the palace, and my father's to him, steadily decreased. I think my father had come to realize that the Swami's pretensions to great occult powers were somewhat over-rated, and his old scepticism had begun to reassert itself. But by then it was too late.

The Jammu and Kashmir State Forces, comprising nine infantry battalions, the bodyguard cavalry and two mountain batteries, were strung out in penny packets along the hundreds of miles of State borders from Gilgit down to Mirpur. The forces had an enviable military record at home and abroad ever since they were raised a century earlier by Maharaja Gulab Singh during the founding of the State; its Hindu and Muslim troops, including Dogras from Kangra and Gurkhas from Nepal, had shown exemplary courage and comradeship for many decades. But now a new factor had entered, the grim significance of which was not realized by my father or his advisers. The virus of communalism that was raging in the subcontinent could not have left our State uncontaminated. The main recruiting area for the Muslim component of the forces had been the provinces of Mirpur and Poonch, from where several martial classes of Rajput Muslims also provided thousands of recruits for the Indian Army. With the creation of Pakistan all those regiments opted for that country, and the people of these areas, directly contiguous to the new nation and bound to West Pakistan by bonds of religion and family relationship, were naturally swayed against my father despite their traditional loyalty. Thus not only were the State Forces dangerously over-extended, but one-third had in effect switched loyalty to the other side and were waiting for the opportunity to defect. Added to this was the discontent amounting to revolt in some areas contiguous to Pakistan, and Jinnah's single-minded determination that Kashmir was to be the fairest jewel in the crown of the new State that he had virtually single-handedly carved out of the subcontinent. All that was required to ignite the entire situation was a flaming brand. This came with brutal swiftness in the shape of the infamous tribal invasion of Jammu and Kashmir.

The storm finally broke on 22 October that year. It was Dussehra and, incredibly enough, the annual Darbar was actually held in Srinagar as usual. By then serious trouble had broken out on the borders, and my father had sent Brigadier Rajinder Singh Jamwal, who had taken over from General Scott as Chief of the Army Staff of the Jammu and Kashmir State Forces, to go to Uri and fight 'to the last man and the last bullet'. I happened to be present when my father called him and told him in Dogri that the situation was very serious and that he was to fight the invaders to the finish. Rajinder Singh was a man of few words, and I recall how after getting his instructions he saluted my father and mother, smiled at me, and briskly marched out of the room. What happened next is part of military history, a saga of courage and dedication that will rank with any in the world.

Rajinder Singh left for the border of Uri-Domel on the night of 22 October and, in a series of incredibly brave last-ditch manoeuvres, succeeded in holding the advancing hordes back for three vital days before they could reach Baramulla, thus gaining the critical time that was required for the Instrument of Accession to be signed and the Indian Army to begin arriving by air at Srinagar. Hopelessly outnumbered, his Muslim officers and troops turning hostile, the critically wounded Brigadier insisted that he be left on the roadside with only a revolver in his hand, as he had promised my father that the enemy would advance only over his dead body. He was the first gallantry award winner of independent India, being awarded the Maha Vir Chakra posthumously for his outstanding act of bravery and sacrifice. Just over a hundred years earlier General Zorawar Singh had created military history by the brilliant Dogra campaigns in Central Asia, and Brigadier Rajinder Singh closed the century of Dogra rule by another heroic act. A simple but moving memorial now stands at the spot on the road to Uri where he took his last stand and fell to enemy bullets after his own were expended, as commanded by his ruler. Meanwhile another senior officer, Brigadier Ghansara Singh, Governor of the Northern Province of Gilgit, had been taken prisoner by the Gilgit Scouts which, under their British Commandant, Major Brown, had switched loyalty to Pakistan.

Of all this we learnt only much later. On that fateful day I was left virtually alone in the palace while my father and members of the staff were attending the Darbar in the beautiful hall at the city palace on the Jhelum with its richly decorated papier mâché ceiling. Suddenly the lights went out—the invaders had captured and destroyed the only power house at that time which was at Mahura on the main road from Domel to Srinagar along which the invasion was proceeding. In the pitch darkness I sat quite alone in my room on the wheel chair. After a few minutes the eerie silence was broken by the sudden, blood-chilling howl of jackals. Weirdly the cacophony rose and fell, then rose again into a mad crescendo. Death and destruction were fast approaching Srinagar, our smug world had collapsed around us, the wheels of destiny had turned full circle. Suddenly there was a flurry of activity. The subsequent events are a jumble in my mind—the servants frantically rushing around with petromax lamps, my father's precipitate return from the Darbar ashen-faced and grim, V. P. Menon's dramatic air dash to Srinagar to persuade him to move down to Jammu, my father's reluctance but final agreement in view of Menon's insistence, and then the long, nightmarish exodus from Srinagar late at night on the 27th.

My mother showed tremendous composure and presence of mind in that hour of peril. The wife of the Deputy Prime Minister, Mrs Batra, had several fainting fits, but my mother collected all the ladies and families of the staff, organized their feeding and put them up for the night, aided by my Australian nurse, Mrs Stewart. Finally the convoy began to move. My father drove his own car with Victor Rosenthal at his side and two staff officers with loaded revolvers in the back seat. My mother followed with the ladies in several cars. I was in no position to get into a car because of the heavy plaster cast, so my wheel chair was lifted and placed in the back of one of the station wagons that my father used for his shikar expeditions. It was bitterly cold as the convoy pulled out of the palace in the early hours of the morning. The raiders were pouring in from across the border, pillaging, looting and raping as they came, and there were rumours that the road to Jammu had been cut and that we were likely to be ambushed on the way. There was no armed escort except for the palace guards, and as we left we had only our faith in God to sustain us. The journey was interminable, with numerous stops *en route*. I fell into an uneasy sleep, half hoping that I would awake to the comfort and security of my room and that the whole sequence was only a bad dream. But each time I awoke the night was colder and darker, and suddenly my hip started throbbing with pain.

All through that dreadful night we drove, slowly, haltingly, as if reluctant to leave the beautiful valley that our ancestors had ruled for generations. Our convoy crawled over the 9,000 ft Banihal Pass just as first light was beginning to break. When we stopped at Kud, a small settlement sixty miles from Jammu, we saw that a cream coloured car had joined our grim procession. It was Swami Sant Dev, whose miraculous powers did not include the capacity to face the raiders. Victor told me later that throughout the journey my father spoke not a word as he drove. When the next evening he finally reached Jammu and pulled up at the palace he uttered but one sentence—'We have lost Kashmir'.

The weeks that followed are still blurred in my memory. I continued to be immobilized as my hip had shown no sign of improvement, and as month followed month I began to get the terrible feeling that perhaps I would never be able to walk again and would remain an invalid throughout my life. I continued to maintain a cheerful exterior, which surprised others, but deep inside this gnawing fear grew, and often I would lie awake far into the night with a feeling of helpless anguish. By then the Pakistani invasion was in full swing. The tribals, aided and abetted by Pakistani troops in mufti, had poured into the valley and almost succeeded in reaching Srinagar and capturing the airport. Had they done so, the valley would have been doomed because there was simply not enough time for the Indian Army to move by road. The gallant suicide stand of Rajinder Singh and his men had held the raiders up for three crucial days, thus enabling the historic airlift to get underway. The Indian Air Force and the Indian Army with tremendous courage and sacrifice were able first to save Srinagar by the narrowest of margins and then launch a counter-offensive.

A Brigadier Paranjpye was commanding the troops in the Jammu sector under the overall command of General Kulwant Singh, and they came quite often to the palace for a drink and discussion with my father. The then ruler of Patiala, Maharaja Yadavendra Singh, also came with a contingent of the Patiala State forces. He was a fine figure of a man, standing six foot six in his turban, but his mien was gentle and he talked in a pleasant voice which belied his formidable exterior. Phrases like 'l of c' (line of communications), 'H.Q.' (Headquarters), 'P.O.L.' (petrol, oil and lubricants), 'C.O.' (Commanding Officer) began to figure increasingly in the conversation. In the Jammu Province the invaders had made important advances; Mirpur, Bhimbar, Rajouri, all border towns of considerable importance, had fallen, and the crucial town of Poonch had been besieged. Although maintaining outward composure, with news of each reverse my father would wince inwardly as if something had died within him. Only on one occasion, when the twin villages of Deva and Batala which were the home of the Chib and Bhau Rajputs fell, did I see a hint of tears in his eyes.

My mother was very active in refugee relief work. Thousands of men, women and children were pouring into Jammu from the areas under Pakistani occupation, and were given food and shelter in a number of refugee camps in and around the city. These, of course, were the lucky ones. Many thousands perished in the butchery and carnage that accompanied the invasion. There was hardly any family which did not lose half or more of its members in the slaughter, and there were numerous cases of only a single child escaping from a joint family. The misery and suffering were unspeakable, and it was at this moment that my mother's courage and her surprisingly strong organizing capacity was revealed. She would often spend the entire day visiting one camp after another, distributing rations and clothes deep into the night. Often she financed and arranged for simple marriages among the refugees, thus bringing some comfort to the shattered families. Indeed she and her band of workers began to be looked upon as angels of mercy, and to this day there are people who recall the tremendous services she rendered to the destitute and terrified refugees at that time. She also organized a women's volunteer force known as the Maharani Seva Dal, and arranged for them to receive paramilitary training from the army authorities. This was the second time she had thrown herself into relief work, the first being in the War Aid Committee during the Second World War, and it seems that this sort of situation brought out the best in her. In contrast, my father rarely left the palace during those days, although my mother did succeed in dragging him on rare occasions to a refugee camp or a Seva Dal function.

By far the most memorable encounter for me during that period was my first meeting with Jawaharlal Nehru. On his first visit to Jammu he had come briefly to the palace but been very busy, and I complained to my father of not having had the opportunity to meet him. The second time he came my father brought him into my room. It was a great moment to meet the man who had become something of an idol for me. As he entered the room I was struck by two things, the agility with which he moved and that unforgettable smile—pensive yet intensely human. My father introduced me by saying 'Tiger is your great admirer'. He asked how I was, and good-naturedly apologized for not having seen me on his first visit. I was too overwhelmed to say anything very intelligent and simply asked him to autograph my copy of his *Autobiography*. He was in the room for barely three minutes, but the moment remains with me.

Sardar Vallabhbhai Patel also visited Jammu on two or three occasions, accompanied by the Secretary for States, V. P. Menon, and in fact it was he who realized that at the rate I was going there was little chance of my ever recovering. He suggested that I be sent to America for treatment, a proposition that at first was violently opposed by my mother. The Sardar was particularly considerate to and friendly with my parents, and finally he and my father were able to persuade my mother to agree. I welcomed the idea, not only because of the hope it held out for my recovery but also because I was finding the whole atmosphere of tension and conflict extremely depressing and yearned to get away.

The Sardar arranged for a special foreign exchange allocation to meet the expenses for my treatment, and it was decided that I would be accompanied by Brigadier N. S. Rawat, a senior Gurkha officer of the State Forces, and Captain Ranjit Singh, one of my father's aides. Victor Rosenthal, being my father's only friend with contacts in the U.S.A., was asked to initiate inquiries and make arrangements for my admission to a suitable hospital in New York. He was able to do this within a few weeks, and finally the date for my departure from Jammu was fixed in the last week of December. Before I left I asked my mother to distribute all my clothes to refugee children. I had grown out of them while in bed and there didn't seem much point in storing them. I left the palace in the evening because of an auspicious timing, and spent the night in the Satwari Cantonment. Before I left, my mother bade me a tearful farewell; my father was also moved but, of course, kept up a stern facade. Although none of us said so, at the back of our minds was an unspoken dread that we may never see each other again.

The chartered DC-3 took almost a whole day to reach Bombay, and two days later on 29 December the three of us—

Rawat, Ranjit and I—boarded a four-engined T.W.A. Skymaster at Bombay airport. I was driven up to the plane in an ambulance, then lifted into the cabin and placed on two seats with several cushions in between to make up a sort of bed. I remember the curious feeling I had as I was being lifted up the steps; sorrow at leaving my country mingled with a sense of excitement at the unknown journey that stretched out before me. The American crew entered the plane soon after, and the Captain came and asked how I was. The plane took off around five in the evening. Our first stop was somewhere in the Gulf area, and the second in Cairo. From there we flew on to Rome where we landed in a blinding rainstorm. Passengers kept getting in and out, but I was quite unable to move and watched the proceedings from my window. We then flew on to Paris, which we reached after dark, the city an illuminated carpet from the air. From Paris we went on to Shannon in Ireland and from there started the Atlantic crossing which, in those days of propeller aircraft, seemed interminable. The shortest route between two land points was from Shannon to Gander in Newfoundland, and even that took the better part of a whole day. We reached Gander towards sundown, and from there made the final hop to New York.

New York had just had its heaviest snowfall in sixty years, and when we landed late in the evening of 31 December there was over two feet of snow in the city. The doors of our plane opened to let in piercingly cold winds. Victor Rosenthal's representative, a Mrs Toole, was at the airport, and after long formalities were completed an ambulance drove up to the plane, a stretcher brought into the cabin, and I was lifted on to it and carried down into the vehicle. A news photographer was there to record my arrival. Ranjit sat with me, Mrs Toole in front with the driver, and we started off on our long drive into town, siren screeching and the snowfall reducing visibility to a few feet.

Thus it was, as 1947 came to its close, that I found myself half a world away from home in a strange country, seeing snowfall for the first time in my life, since I had never spent a winter in Kashmir. As the ambulance sped towards its destination my mind went back over the experiences of my life so far—my childhood, school and the frenetic events of the last two years.

Would I ever see India again, or my parents? What would become of the State of which I was born heir apparent? Would I ever be able to walk again, or was I condemned to be a helpless cripple for the rest of my life? At that moment the prospect appeared to be one of unrelieved gloom. The only thing that kept me going was the mantra for worship of the Goddess that my mother had taught me several years earlier, and which she had earnestly enjoined upon me never to forget in times of need and distress. She had also given me a small picture of Durga *Singhavahini*, the eight-armed Goddess riding on a lion, a powerful symbolic representation of the divine astride material power. Throughout the long flight I had clutched it, and in that hour of despondence it provided my only comfort.

We were now entering New York, and I got the first glimpse of the great city that was to be my home for much longer than I realized, its huge buildings thrusting up into the raging blizzard, their tops lost in the swirling snow. After a drive of almost two hours the ambulance finally pulled up at 321 East 42nd Street where stood the Hospital for Special Surgery. It was late at night and at first the institution seemed to be deserted, but soon someone appeared and I was lifted out of the ambulance into the hospital veranda, and wheeled into a lift. Then, at the fifth floor, the lift stopped and I was wheeled along to the end of the corridor into Room 509. By the time I was transferred to my bed and the ambulance attendants departed it was getting on to midnight. They all had New Year parties to attend, and were understandably in a hurry to get away. Ranjit was in Room 508 next to mine, and he also went to his room for a while. At last I found myself alone, with the picture of Durga and a photograph of my parents on the high table next to my hospital bed. I heard in the distance the sound of a great clock striking the midnight hour. A new year had begun and, for me, a new life.

Chapter 6

America is an astonishing country, even when viewed by a teenage Indian from a hospital room. There was something different about the whole atmosphere, an air of freshness and informality. To begin with, the nurses, doctors and ward attendants looked at me with frank curiosity, as a photograph of my arrival had been published in the *Daily Mirror*. All sorts of odd stories had evidently been circulating about Indian princes, and so when they heard I had checked into the hospital they probably expected to see some kind of freak. They were all very friendly and seemed pleasantly surprised that I could speak English. In fact I spoke it better than most of them, and it was several weeks before my ear became accustomed to the curious American accent and what struck me as quaint usages, such as 'crackers' for biscuits and 'cookies' for pastries.

I was visited by Dr Philip D. Wilson the day after I arrived. Dr Wilson was one of the most distinguished orthopaedic surgeons in America at the time, and, like some of the really great doctors, his very presence infused a sense of confidence in the patient. He asked me several questions and then gave directions that the plaster cast on my leg should be removed and the usual battery of X-ray, blood and other tests taken so that he could decide on a line of treatment. The cast was replaced by an eight-pound traction on my foot to keep the joint in position. After months in the cast my knee and ankle were quite stiff, and I could not move the hip joint at all. To help loosen the joints I was taken every morning to a heated swimming pool in the hospital which specialized in treating muscular atrophy. Getting into the water for the first time after six months was a pleasant experience, and I found I could move around with ease. Gradually I started putting weight on my legs, and a few days later was allowed to stand in my room with the help of crutches, an extraordinary feeling after lying flat for half a year.

Soon after arrival we bought a radio, and I was fascinated by the large number of stations available and also the fact that the programmes went on for eighteen hours, from six in the morning to midnight. There was a bewildering array of music, plays, talks, news and sports commentaries, from which I soon chose my favourite items, mainly pop music. In fact music once again played a vital role in a very trying period of my life, and although it was poles apart from Indian classical music I discovered that in whatever form, music has the unique capacity to soothe the mind and prevent it from slipping into depression or worse. Later, a television set was also acquired, which added a new dimension to my life in hospital. There were two additional attractions. One was a weekly film show on the third floor for patients who were in a position to be moved, and the second a solarium on the sixth floor—a large, glass-panelled room with many potted plants and trees. It was marvellous to be able to go up there and savour the warmth of the sun without being exposed to the chill fury of the snow.

The unusual friendliness of the Americans became evident soon after I arrived. Not only did the entire hospital staff go out of its way to be pleasant, but several visitors began coming in to see me, including some old Indian contacts and friends of Rosenthal's. Among them were a middle-aged couple, Norris and Dorothy Harkness, and an old India hand, Major John Nethersole and his wife Arlene. Looking back, I find it amazing how much all of them did to befriend me. They would drop in every two or three days, bring sweets and 'cookies', and sit with me for hours chatting away or playing card games and backgammon. The women in particular seemed to develop an almost motherly affection for me. I wonder whether the fact that I was missing my mother so much in some way conveyed itself to them.

Meanwhile the Kashmir dispute in the United Nations was in the front pages of all the U.S. newspapers. The wily Zafarullah Khan, Pakistan's able chief delegate, seemed to our utter exasperation to be getting the better of the Indian delegation led by no less a person than the former Prime Minister of Jammu and Kashmir, Sir Gopalaswami Ayyangar. The latter came to see me in hospital one day, soon after the shocking news of Mahatma Gandhi's assassination on 30 January, a tired and broken man. Another member on the Indian delegation who also visited me was Sheikh Mohammad Abdullah. Having heard so much about him since my childhood I was quite excited by his visit, and, indeed, he was an impressive figure of a man, six foot four and well built. It was curious that a man with whom I was to develop such intimate political relations met me for the first time half a world away from our home State.

Despite all these visits and diversions, the fact remained that I was still a cripple, unable to move on my own, still miserably dependent upon sponge baths and bedpans. After studying my X-rays and bringing in two more specialists for consultation, Dr Wilson decided by the end of January that, while an operation was not immediately necessary, they would try a manipulation and replastering of the hip joint. Thus began the first of a whole series of encounters with anaesthesia. For those who have not undergone this voluntary loss of consciousness there may appear nothing very special about it, but anyone who has will testify that the prospect of going through a belt of unconsciousness is not a pleasant one. However, my main concern was not so much the unconsciousness itself as the horrifying thought that I might awake in the middle of the operation. Each time I would tell the anaesthetist to make quite sure that I remained under until the operation was over.

Two main fears that I had about anaesthesia were that I might lose my mental agility or my virility. Each time I would come round after an operation I would check as to whether I was still mentally alert by—oddly enough—trying to recall the definition of 'optimum population' that I had learnt for my Intermediate economics examination There was no special reason for my having chosen this particular mental formula to test myself, but it is a strange coincidence that a quarter of a century later I was to be involved in framing a National Population Policy. The test for my other fear would have to wait for a few days until the pain of the operation had subsided.

The process of treatment, plastering and replastering went on for several months, as the freezing New York winter gradually changed into a hot and sticky summer. I had picked up several hit tunes from the radio, but, in addition, I decided to start taking chess lessons and also re-establishing some contact with the academic world. For chess it was arranged that a man called Boris Siff would come thrice a week. He was very good, and introduced me to new dimensions of this fascinating game.

On the other three days a Mr J. E. Brown (I am sure it was an assumed name, because he spoke with a gentle Central European accent) came from Columbia University to give lessons in economics and political science. He was very leftist in his leanings, and a strong critic of the American capitalist system. It was he who first put before me the theory that American munition manufacturers were actively involved in fomenting wars in various places in South America so that their wares could retain their regular and highly lucrative markets.

As the weeks went by I began to realize that my hopes for an early recovery in America were steadily receding. Meanwhile events were moving fast in Jammu and Kashmir. The tribal invasion had developed into virtually a full-scale war, in which the State Forces and the Indian Army were pitted against the tribals and regular units of the Pakistan Army. My father's letters gave me news of the military campaign, from which I gathered that the Indian offensive was gradually succeeding in pushing the invaders back. Town after town was recaptured, but the process was much slower than we had hoped. Politically also, changes were taking place. Mehr Chand Mahajan had left the State, and my father had set up by proclamation an Emergency Administration headed by Sheikh Abdullah. It was quite clear that this had been done at the instance of Jawaharlal Nehru. Indeed, after the Instrument of Accession was signed and the Indian Army moved in, effective authority had in fact passed from my father's hands, though this was not formalized till later.

The Constitution of India was also being framed at the time. In one of his letters my father wrote prophetically, 'I have no idea of how the new Constitution for India is going to affect us beyond the fact that before very long the Rulers will disappear even in name'. In hospital I looked eagerly through the gigantic daily newspapers for news about India, which was sparse and unsatisfactory. The size of American newspapers never ceased to surprise me, specially the Sunday edition of the *New York Times* which, with its numerous supplements, sometimes reached two hundred pages. I was an avid reader, and lapped up a great deal of information and new ideas.

Although immobilized due to the plaster cast, I found both my body and mind growing almost visibly, the former mainly because I drank vast quantities of delicious homogenized milk. On my seventeenth birthday we had a party in the hospital, and my photograph receiving the birthday cake from Dr Wilson flanked by my two nurses and a Chinese fellow patient, Tai Loong Yang, was published next day on the front page of the *New York Times*. The report ended with the following cryptic paragraph: 'Sir Hari Singh has recently liberalized his formerly despotic government and his son is said to have democratic ideas. The youth declined, however, to discuss any political matter.' I also did my first radio broadcast at about this time on a show called 'Junior Reporter' in which teenagers discussed the topic of the day. In my intervention I said I hoped India would soon assume her rightful place among the great nations of the world, that the responsibility of the future world rested on its youth and that we should all strive for better understanding and brotherly love.

Looking back I realize that the American interlude with its totally new environment provided my mind the opportunity to break out of the feudal trappings of the State and absorb a different and much freer atmosphere. 1948 happened to be an election year, and I followed closely in the press and on TV the astonishing American electoral process, including the Democratic Convention at which Hubert Humphry, Harold Stassen and others made an unsuccessful bid to wrest the nomination from President Truman, and the Republican session where Thomas Dewey was chosen in an atmosphere of euphoria. The cheer-leaders, bands and the *tamasha* at these conventions amazed me, but Mr Brown kept pointing out that behind all the noise and festivities there were hard-nosed professionals representing various interests who would see to it that neither party strayed too far from the path of capitalism and 'free enterprise'.

The letters I wrote to my father at the time were full of the Security Council proceedings regarding the Kashmir dispute, and questions about the progress of the military campaign in the State. In reply he would give me the salient developments, mentioning specially the recapture of any important town. Surprisingly, he still kept up his interest in the Bombay races as his letters often complained about his trainer and the poor performance of his horses. I also wrote regularly to my mother in

Hindi, and she would reply on her salmon pink paper in an almost undecipherable hand. She had never undergone any formal education, and though in conversation and even in public speeches she was impressive, writing was never her strong point.

Despite the prolonged immobilization of my hip in plaster, now almost a full year since it had first been treated by Dr Mirajkar in Srinagar, the joint did not respond to Dr Wilson's satisfaction. On 6 July Dr Wilson came into my room and said that he had decided to operate upon me the next morning. Evidently the decision had been taken several days earlier, and my father's permission secured, but it had been kept from me so that I would not worry and fret. The conspiracy of silence worked well, because I only had a few hours to worry before the big event. The operation that took place the next morning was much more serious that any of us had realized. It involved a semi-circular incision twelve inches long, and permanent immobilization of the hip joint through a bone graft and a six-inch metal pin. I was on the operation table for several hours, and regained consciousness only late that afternoon. The next five days were the worst that I could ever recall till then. My body was shaken by convulsions caused by the severance of the major hip muscle, the pain was excruciating and with each convulsion the whole bed would shake. I had to have sixteen injections every day—eight of pencillin and eight for the pain, and was given five 250 c.c. bottles of blood by transfusion. If heaven or hell actually exist outside one's consciousness, I think I got a fair approximation of the latter during those terrible post-operation days.

Ranjit was in my room almost round the clock and, along with the three nurses who came for eight hours each, did everything possible to try and lessen my agony. The anaesthesia left a peculiarly unpleasant aftertaste, and I totally lost my appetite. In any case, I was hardly in a position to eat, and was on liquid diet for a week. It was only casually, a few days later, that one of the nurses broke the news of the pin in my hip. My heart sank; I realized then that I would never be able to walk normally again, and indeed at that point was virtually convinced that I would never be able to walk at all. My faith and such courage as I had been able to muster were sorely tested, and it was only with a supreme effort of will that I resisted breaking down completely.

Dr Wilson, though, was happy with the operation, and assured me that all would be well in another few months. The fact that to this day I can play tennis and undertake prolonged election tours without any discomfort is a tribute to the professional competence of that great surgeon, one of the most distinguished names in the history of orthopaedic surgery. He would sit by my bed, telling me in a fatherly manner how he had used the bone graft from the newly developed bone bank. He also gave me a replica of the special alloy pin that had been inserted in my hip, which remains with me as a rather gruesome reminder of one of the critical periods in my life.

I was young and, evidently, stronger than I realized. My first post-operative letter to my father was scrawled on 14 July, exactly a week after the operation, and on the 19th I wrote him a seven-page epistle. By then my wound had begun healing, and the twenty-five stitches removed. Slowly, very slowly, the pain subsided and my appetite returned. The vibrancy of youth began to assert itself, and I think the fact that I refused to go into a psychological depression made a big difference. Everyone we knew in New York and in the hospital was most kind and affectionate; during the several months I had been there I had made more friends than I realized, and my room was always full of flowers and 'candy'. As I rigorously concealed my pain and distress as far as possible, I got the reputation of being brave and courageous. Even my father in his next letter wrote: 'You have patiently gone through a very great deal of trouble bravely. . . . I am really proud of you.' High praise, indeed, coming from him.

For two months I wore a heavy plaster cast and then in September, after Dr Wilson returned from his summer holiday, he had the cast removed and a smaller, lighter one applied. The operation wound had healed perfectly by then, and I remember vividly the first day that I was allowed to stand after months in bed. The sensation was so strange that I felt giddy and almost fell. I was made to stand for longer periods each day, and to take the first tentative steps in the hospital room, virtually relearning how to walk. On 24 September I was well enough to be able to leave the hospital for the first time in nine months and go for a short drive in an open car. I was taken down in a wheel chair and lifted into the car. Never will I forget the sheer joy of that first drive in New York city. Those tremendous buildings soaring up into the air were quite overwhelming, and I had a peculiar sensation of sensory distortion, of being suddenly transported into a new dimension of space and time. After the long, dreary months in hospital the very thought that I could once again move around was delicious.

During these months my chess and political science sessions continued. I also started (but did not complete) a Pelmanism course by correspondence with Bombay, and a Speed-writing course which involved a method of condensed writing without using any shorthand. I watched television ardently, followed closely the Presidential campaign with Truman and Dewey criss-crossing the country making unending speeches. Although everyone seemed to think that the rather square Dewey was certain to win, it struck me that Truman had more vigour and fight in him. I read a great deal, books on contemporary politics as well as novels.

News from home remained disturbing. My father was very guarded in his letters—he had warned us that censorship was in the air—but it was clear that although he had been forced by Jawaharlal Nehru to hand over effective power to Sheikh Abdullah, who headed the 'Emergency Administration' and later became Prime Minister, their mutual relations were tense and antagonistic. For example, in a letter dated 18 September he wrote:

> The present government was very eager to take over the State Forces under their control. I strongly opposed this move on their part as the Army was a reserved subject. The matter was referred to the States Ministry and I was invited by Sardar Patel to come to Delhi for discussions. After a long discussion, contrary to the wishes of Sheikh Abdullah it was decided that the control of the State Forces should be handed over to the Indian Army for the time being. I would continue to be Commander-in-Chief.

My mobility gradually increased. I began walking with crutches and, accompanied by Ranjit, would walk out of the hospital and down to the corner drug store for ice cream. I was also able to walk into the ground floor of the *Daily News* building on 42nd Street where a huge revolving globe was set in a room full of time charts, news flashes and maps. It was a fascinating place, and I became a regular visitor, spending as many as two or three hours there at a stretch. I had bought myself a couple of suits, ties and a felt hat, and was able to send back some photographs to my parents who were delighted to see how I had grown. Meanwhile, the American Presidential election moved to its climax. Truman scored a stunning upset victory against Dewey, and I won ten dollars off one of the doctors who had given me ten to one against the incumbent President winning.

Although I was gradually improving, I had no illusions about a rapid or complete recovery. In one of my letters to my father I wrote: 'The loosening and strengthening of the muscles of my right leg will be a long slow process, but once I get rid of the plaster permanently it will only be a matter of time. Of course, as you know, my right hip will always be stiff, but after some practice and adjustment in my way of walking and sitting, I do not think that should trouble me too much in life. It will, however, take some time to get used to it.' On 1 November the plaster cast was finally removed, and I was able to restart going into the pool. It was marvellous getting into warm water again, free of that dreadful cast. The competent silver-haired physiotherapist, Miss Helen Clark, who had treated me regularly ever since I entered hospital, gave me underwater massage, as a result of which my right hip and leg gradually began to recover some strength. Dr Wilson was so pleased with my progress that he said I could leave the hospital after a few days and stay in a hotel for another two months so as to be able to come in for physiotherapy and observation by him. So at last on 15 November, exactly 10½ months since I had been carried into the Hospital for Special Surgery, I moved out to the Barclay Hotel, 111 East 48th Street.

In one of his poems Byron suggests that even a prisoner who stays in a single cell long enough becomes attached to the jail. Certainly I had developed a good deal of affection for the hospital—the nurses, the ward staff, the doctors, the fellow patients. Although I suffered physically a great deal there, the atmosphere of affection and consideration helped sustain me. It was in fact quite a wrench to leave, and I was given a royal send off by the entire hospital. Twenty years later, in 1967, I visited New York again as Minister for Tourism and Civil Aviation of the Government of India. The old hospital on 42nd Street had by then been pulled down, and the institution moved in a much more modern and slick building further up town. Dr Wilson, still vigorous and ruddy at eighty, showed me proudly around the gleaming new premises. And yet, somehow, I missed the old.

The Barclay was a spacious and comfortable hotel, not nearly as luxurious as the Waldorf Astoria close by, but well suited to the comparatively limited dollar remittances that my father was allowed to send for my treatment. I particularly enjoyed the luxury of having hot baths again after so many months. To this day I offer up a silent prayer of thanksgiving whenever I sink back into a tub of hot water, and remind myself of how breezily we tend to take the comforts of life for granted until they are denied to us. I was still walking on crutches when I moved into the hotel, but soon started getting around with the help of two canes. Thrice a week I would go to the hospital pool for underwater muscle therapy, while on the other three days Miss Clark would come to the hotel and give me an hour's massage. I was now able to get into taxis and so, along with Rawat and Ranjit, began to take in the sights and sounds and food of New York.

It is truly an incredible city, brighter at night than in the day. Times Square and Broadway were particularly impressive, with thousands of neon signs bursting forth in a permanent burst of light. The huge cinema advertisements fascinated me. *Joan of Arc* had just been released with Ingrid Bergman in the title role, and a gigantic cut-out of her stood higher than any building I had seen in India. We saw this and a number of other movies, including *The Three Musketeers* (with Gene Kelley, Lana Turner, June Allyson and Angela Lansbury), *Enchantment* (David Niven) and *The Paleface* (Bob Hope, Jane Rusell). We were also able to see some plays on Broadway: *Mister Roberts* (with Henry Fonda), Noel Coward's *Private Lives* (with Tallulah Bankhead), *Harvey* (starring a huge, invisible white rabbit) and a comedy *Where's Charley?* (with Ray Bolger).

In addition to movies and plays we went to the top of the

Empire State Building, visited the impressive Rockefeller Centre, shopped at Saks and Macys, admired the famous Rockettes at Radio City Music Hall, witnessed an ice hockey match at Madison Square Garden, cautiously explored the outskirts of Central Park, and discovered a number of excellent eating places including several Indian restaurants and one called The Barberry Room (19 East 52nd Street) with a starry ceiling and the most comfortable chairs in New York. Just being in the city was an exhilarating experience, specially after having been laid up for eighteen months, and those weeks in New York out of hospital were among the happiest in my life. We also took an overnight train journey to Buffalo from where we went to the Canadian border to see the Niagara Falls.

1948 had drawn to an end, and, unlike the previous New Year's eve when I had been driven in an ambulance along deserted streets, this time we celebrated it in Times Square along with half a million New Yorkers. It was bitterly cold, but we were in high spirits when the clock struck twelve and everyone went berserk. Soon thereafter we found ourselves in an Indian restaurant nearby, where the proprietor served us steaming cups of saffron tea. I had heard from several friends, including Gurubachan Singh of our Foreign Service who was posted there, that Washington was a beautiful city, and so I asked my father for permission to visit it for a week. He agreed, and we went there just in time for President Truman's Inaugural Parade on 20 January 1949 for which we got tickets through Gurubachan's good offices. Our Ambassador, Shri Rama Rau (whose brother Sir B. N. Rao had been my father's Prime Minister) was out of town, but his gracious wife, Lady Dhanwanti Rama Rau and their daughter, Mrs Premila Wagle, took good care of us. Washington is indeed a lovely city, specially its public memorials. The classical Greek severity of the Lincoln Memorial, with its huge brooding statue of the greatest President, contrasted sharply with the less formal and more intimate Jefferson Memorial set amidst pools and flowering bushes. We also went to Mount Vernon to visit George Washington's ancestral home. Both the unrelenting glamour of New York and the classic dignity of Washington, so different from each other, made a powerful impression upon me.

Now that I was improving, my parents began wanting me to

come home, and in my letters to my father I began zeroing in on two major questions that would confront me when I returned. The first was my marriage, which I presumed—erroneously as it turned out—would be to Shanti with whom I had begun corresponding from hospital. I wrote to my father saying that while I appreciated their desire for an early marriage, 'I am sorry that I will not be able to see more of Shanti and get to know her a little better before we are married, because I am sure that if two people know each other well there is more chance of a successful marriage than if they are almost strangers.' The second matter was about my future education, and here I was emphatic enough to read my father a virtual sermon:

> I would like you to know that I am very keen to continue my studies up to B.A. and further, that I sincerely believe, and I am sure you do too, that in these rapidly changing times a sound education (in my case subjects such as Economics, Law, etc.) is essential. It is not enough to be born into a high position. I believe that one should be, in these times, capable enough and well enough equipped educationally to make a mark for oneself regardless of any accident of birth in a high position.

Dr Wilson was happy with the way my leg was improving, and it was finally decided that I could leave for home in the first week of February, stopping for a few days each at London and Paris on the return journey. One of the last things we did in New York was to pay a visit to Roosevelt House at Hyde Park along with one of the late President's sons, Eliott Roosevelt and his wife Faye, who had been an actress. We happened to be there on the birth anniversary of the late President, whom I always greatly admired, and we participated in a short memorial service at Hyde Park where he is buried. We were greeted and shown around by his widow, Mrs Eleanor Roosevelt, a woman of great drive and personality. In her column in the *New York World Telegram* on 2 February she recorded our visit and added: 'The young Prince of Kashmir, who has been in this country for an operation, is returning home in a few days to complete his final two years in college. It is his desire later to serve in the United Nations and help to build a peaceful world.'

The return trip to India was in sharp contrast to the grim journey to New York over a year earlier. We boarded a Pan American flight across the Atlantic as Air-India had not yet started plying on this sector. The first stop was London, where we stayed at what had been my father's favourite hotel, the Savoy. Characteristically, he had sent me a meticulously prepared list of clothes I should buy, along with the shops that he himself used to patronize as a young man. Captain and Mrs Wreford came up from the country to see me, and several friends of my father also called. The only trip into the country was to the public school at Charterhouse to meet Shanti's younger brother, Ranbir Singh. The Savoy theatre had just begun showing the celebrated *Mousetrap*, Agatha Christie's all-time theatrical record-breaker, which we saw with interest.

After about a week in London we proceeded to Paris where Victor Rosenthal had laid on a lavish welcome. Two of his brother's grandsons—Jean and Hubert—who were about my age showed me around town, but I was far too shy and self-conscious about my cane to be able to really enjoy myself. The Galeries Lafayette was the most beautifully decorated department store I had ever seen. We visited two famous night-clubs, Lido and Moulin Rouge, and at one restaurant heard Edith Piaf singing. There was some talk of visiting my birthplace in Cannes, but my father was keen that I should not tarry too long on the way home, and so we dropped the idea.

Finally, in the third week of February, we set out on the last lap of the flight back to Bombay. I returned with mixed feelings. I had set out from home over a year ago as a helpless invalid, and though I was now able to get around on my own I had not fully been restored to normal health despite Dr Wilson's assurance that gradually I would be able to play tennis again. In addition, there had been ominous hints in letters from home that politically things were not going well as far as our family was concerned. Besides, the whole question of my forthcoming marriage was in some ways disturbing. It was, therefore, with a sense of foreboding that I awaited the descent of our plane to Bombay late in the evening.

Chapter 7

The welcome I received on my return only heightened my fears. Instead of the enthusiastic greeting that I expected, an old friend of my father, Fatehsingji of Limbdi (Uncle Fatty) met us at the foot of the ladder and said that I must come quietly with him to a car in which my father was waiting for me. Evidently they had received some sort of assassination threat, and were afraid of giving undue publicity to my arrival. Once in the car we sped off to 19 Nepean Road which my father had bought some years ago from the Modys. It was very odd meeting my father after so long under such circumstances. When we reached the house, my mother, who was recovering from an operation for piles, rose to greet me and clasped me to her bosom. I was home at last after almost fourteen months, but all I felt after the long flight was tiredness and a strange sense of let down. As soon as we got to the house my father, who was always impatient when it came to something in which he was interested, asked me to unpack and show him the things we had bought under his instructions in America. Those included shaving equipment, cigarette lighters and other mechanical and electrical gadgets that gave him great pleasure because of his curiously inventive turn of mind. I had also bought some perfume and other cosmetics for my mother. From the next morning, however, grim and disturbing tidings began coming in. I learnt that the political situation in the State was very bad, and that there was tremendous tension between my father, who was now virtually a constitutional figurehead, and Sheikh Abdullah who, with Jawaharlal Nehru's support, was heading the State administration as Prime Minister and wielded executive power. The old bitterness between the Dogras and the Kashmiris had surfaced again with the dramatic reversal of their roles after a century of Dogra rule. Although, as I have recorded, my father was not really interested in exercising political or administrative power, he was understandably jealous of his authority and prestige as Maharaja, and deeply resented the manner in which Jawaharlal had made his handing over power to Sheikh Abdullah a virtual condition for extending military aid to save the State from Pakistani occupation. This marked for me the beginning of a deep conflict between loyalty to my father who, I realized, represented a system and mode of thought rapidly becoming obsolete in free India, and regard for Jawaharlal Nehru, who was my political guru and whom I greatly admired.

Another startling development that I learnt of soon after I returned was that my father had decided that Shanti was not really a suitable bride for me, and wanted to break off the engagement. There was really no justification for this, but then the engagement itself had been an absurd move based on one of my father's mood-cum-impulse decisions that defied rational analysis. I had no say in the engagement decision, and although this time the matter was put to me for my views I realized that my father had already made up his mind and there was little I could do to change it. I had not seen Shanti at all since her visit to Srinagar several years earlier, and although we had corresponded while I was in America I had to admit that I did not have any emotional involvement. So the decision was taken, grossly unfair as it was to the Ratlam family, and an emissary sent to Ratlam for the purpose. Incredibly, my father had the nerve to ask for the return of the jewellery he had given Shanti at the betrothal ceremony. By then the old Maharaja Sajjan Singhji was dead and Shanti's brother, Lokeshwar Singh, was the titular ruler. It gave me some grim satisfaction when he bluntly refused to return the jewellery.

I learnt soon what lay behind these strange developments. During the Hyderabad operations a military contingent had been sent to India as a gesture of goodwill and friendship by the Rana Prime Minister of Nepal, Maharaja Mohun Shumsher, under the command of his elder son General Sarada Shumsher. At a wedding in Delhi between the Raja of Khetri and Princess Bhuvan, daughter of the Nepalese Ambassador General Singha who was a younger brother of the Prime Minister, my parents had met General Sarada, his wife and their children. The eldest child was a girl, Yasho Rajya Lakshmi, then barely twelve years old. Apparently it was Sardar Patel who hinted that a matrimonial alliance between the ruling families of Kashmir and Nepal might have valuable political implications for the new India that he was so painstakingly building up. At least this is what I was told, though I never was able to get the matter verified. In any case, the timely appearance of the Nepalese princess gave my father the impetus he needed to break the Ratlam engagement, although he had already begun cooling off towards it after Maharaja Sajjan Singhji passed away. It was a sign of the changing times that my parents wanted me to meet the princess and decide whether I liked her or not. On our way back from Bombay to Jammu, therefore, we stopped off at Delhi and the meeting was arranged at Faridkot House over lunch. We were met by General and Rani Sarada and the princess who, although only a couple of months over twelve, was very pretty even at that age. We ate in silence while our parents—mainly the mothers—talked incessantly. After the meal we got into the car and I said 'yes'. The princess was, I learnt later, not even asked about her reactions. It was under these unlikely circumstances that our marriage was decided upon. Some people say that we have both been lucky, but I can speak only for myself.

We received a tumultuous reception when our chartered DC-3 touched down at Jammu. The whole town was out to greet us on my return from America after a prolonged absence, and the people of Jammu took the opportunity to reiterate their loyalty and affection for the Dogra ruling family. I had never seen so many people together at the same time; it took us the better part of three hours to get from the airport to the palace via the Raghunath Temple. In the days that followed I was whisked around to refugee camps where it was heart-rending to see the plight of thousands who had barely survived the tribal invasion. Many had lost all the members of their family in the gruesome massacres that took place in Muzaffarabad, Mirpur, Rajouri, Bhimbar and other towns. All had lost their homes and property, and were huddled together in a number of camps where they were totally dependent upon doles. My mother had done remarkable work in refugee relief, throwing herself heart and soul into the work, collecting money and spending many hundreds of thousands of rupees of her own. I went with her to most of the camps, and the refugees would fall at her feet with tears of gratitude. My father also came sometimes, but he was averse to displays of emotion and the visits with him had a rather more formal atmosphere.

A great deal had changed since I had left. The Indian Army

was now an important presence. My parents had excellent relations with senior army officers, and they would often come over in the evenings for a chat over drinks. The overall Army Commander for the J & K sector was General Kulwant Singh, while General Thimayya was the Divisional Commander in the Valley who had been responsible for the major counter-offensive that drove the invaders back until a ceasefire was declared under the United Nations auspices on 1 January 1949. There was also a distinct change in the sort of people who came to the palace. While in the past the visitors were confined to a small and carefully selected group of courtiers and advisers, now a large number of new faces were in and out, evidently part of a last, desperate attempt to win adherents in the political struggle against Sheikh Abdullah.

The burning issue at that time was the proposal for a plebiscite. While accepting the Instrument of Accession signed by my father, Lord Mountbatten as Governor-General of India had said in his reply:

> In the special circumstances mentioned by Your Highness, my Government have decided to accept the accession of Kashmir State to the Dominion of India. Consistently with their policy that, in the case of any State where the issue of accession has been the subject of dispute, the question of accession should be decided in accordance with the wishes of the people of the State, it is my Government's wish that as soon as law and order have been restored in Kashmir and its soil cleared of the invaders, the question of the State's accession should be settled by a reference to the people.

This 'offer' became the main source of trouble and difficulty later, and Jawaharlal Nehru came in for a good deal of criticism for having unnecessarily internationalized the Kashmir dispute. Indeed, my father's earlier reluctance to accede to either dominion, and Jawaharlal's later offer of a plebiscite, are both matters upon which future historians will doubtless lavish generous criticism. And yet the trouble with historians is that they are seldom 'there', and therefore find it difficult to appreciate the intangible parallelogram of forces and factors at play at any given point in time and space. I have already had something to say about my father's dilemma. Jawaharlal also must have found himself in a difficult predicament. On the one hand was his deep attachment to Kashmir, the land of his forebears, and the compulsions of newly resurgent Indian nationalism. On the other was his genuine idealism, his total commitment to democracy, and his personal affection for Sheikh Abdullah. No doubt Sardar Patel would have handled the situation differently as, in fact, he did in Hyderabad and Junagadh. And yet when a nation has the fortune of a genuine visionary as its leader, it sometimes has to pay the price of idealism itself.

Plebiscite being the watchword at that time, this became the trump card in the hands of Sheikh Abdullah. As the man who was supposed to win the plebiscite for India, he could virtually demand his pound of flesh. And this he did repeatedly; not only by assuming power, but also by pursuing my father relentlessly. He made accusations against him for having organized the massacre of Muslims in Jammu through the RSS during the partition holocaust, accusations that have never been substantiated. We were told that he went to the extent of telling Jawaharlal that my father distributed sweets when Gandhiji was assassinated. Jawaharlal and my father were in any case hardly the best of friends, and Sheikh Abdullah was able to convince him that as long as my father remained in the State he could not possibly win the plebiscite. The political tension between them was becoming intolerable, and it was only a matter of time before one of them had to be eased out.

It will be useful here to recount some of the background of this conflict between two strong personalities—my father and Sheikh Abdullah. Even before the invasion, an effort had been made through my uncle Thakur Nachint Chand to arrange a meeting between them. Indeed, on 26 September 1947 Sheikh Abdullah had written to my father from jail a letter which Mehr Chand Mahajan described as a 'qualified apology'. It ran as follows:

May it Please Your Highness,

It is after about one and half year's incarceration that—as long wished—I had an opportunity of having detailed talks with Thakur Nachint Chand Ji. What unfortunate things happened during this period in the State I need not mention. But this is now realised by every well-wisher of the State that many of the regrettable happenings of the past have mainly been due to the misunderstandings which appear now to have deliberately been created by interested people in order to achieve their own ends. R. B. Ramchandra Kak, the ex-Prime Minister through his mischievous methods and masterly manoeuvrings brought these misunderstandings to a climax and succeeded in his attempt, though temporarily, to a certain extent. He painted me and my organization in the darkest colours and in everything that we did or attempted to do to bring Your Highness and your people closer; base and selfish motives were attributed to me. But God be thanked that all these enemies of Your Highness and the State stand exposed today.

In spite of what has happened in the past, I assure Your Highness that myself and my party have never harboured any sentiment of disloyalty towards Your Highness' person, throne or dynasty. The development of this beautiful country and the betterment of its people is our common aim and interest and I assure Your Highness the fullest and loyal support of myself and my organization. Not only this but I assure Your Highness, that any party, within or without the State which may attempt to create any impediments in our efforts to gain our goal, will be treated as our enemy and will be treated as such.

In order to achieve the common aim set forth above, mutual trust and confidence must be the mainstay. Without this it would not be possible to face successfully the great difficulties that beset our State on all sides at present.

Before I close this letter I beg to assure Your Highness once again of my steadfast loyalty and pray that God under Your Highness' aegis bring such an era of Peace, Prosperity and Good Government that it may be second to none and be an ideal for others to copy.

Your Highness' Most Obedient Subject
S. M. Abdullah

This hopeful letter led to Sheikh Abdullah's release from jail and was followed up by a meeting between him and my father at Gulab Bhavan at which, I was told, he even offered my father a gold sovereign in the traditional court manner. But evidently nothing concrete emerged, and soon the whole situation underwent a drastic change as a result of the Pakistani invasion and the accession. Thereafter the die was cast against my father, and the Sheikh rode the crest of a popular wave with the full support of Jawaharlal Nehru. Apart from his close personal friendship with the Sheikh, Jawaharlal was genuinely convinced that in the peculiar circumstances of Kashmir with its Muslim majority it was absolutely essential both for national and international reasons for Sheikh Abdullah to be fully involved in the government of the State. Thus in a letter to my father dated 13 November 1947 he wrote:

> As I pointed out to you, the only person who can deliver the goods in Kashmir is Sheikh Abdullah. He is obviously the leading popular personality in Kashmir. The way he has risen to grapple with the crisis has shown the nature of the man. I have a high opinion of his integrity and his general balance of mind. He has striven hard and succeeded very largely in keeping communal peace. He may make any number of mistakes in minor matters, but I think he is likely to be right in regard to major decisions.
>
> But the real point is that no satisfactory way out can be found in Kashmir except through Sheikh Abdullah. If that is so, full confidence must be placed in him. There is no half-way house between full confidence and a half and half affair which has little advantage and many disadvantages. Even if a risk has to be taken in giving this full confidence, that risk has to be taken. There is no other way as far as I can see it both from the short-term point of view and the long-term one. Sheikh Abdullah is earnestly desirous of co-operating and is amenable to any reasonable argument. I would suggest to you to keep in close personal touch with him and deal with him directly and not through intermediaries.

I have often wondered how very different the politics of the subcontinent would have been if some understanding could have been reached between the Sheikh and my father, both proud and authoritarian men. Looking through some of the papers that came to me after my father passed away, which includes the letter I have just quoted, it is clear that after power had been handed over to Sheikh Abdullah the tension between him and my father continued unabated. Evidently Jawaharlal did make an attempt to strike some sort of an equation with my father, as there are a number of long letters by him during the period 1947–8 in which he tried to make certain points about the prevalent political situation. But the tension between my father and the Sheikh came in the way of any real understanding developing. There were several causes of conflict. One was the future of the Jammu and Kashmir State Forces, of which my father remained Commander-in-Chief. Sheikh Abdullah evidently demanded that with his appointment as Prime Minister the administrative control of the State Forces should also be transferred to him. When this was strenuously objected to, he suggested that their operational and administrative control be handed over to the Indian Army. While operationally the forces had been under Indian Army control ever since the invasion, my father was evidently reluctant that their entity should be entirely merged with the Indian Army. In reply to a memorandum from Sheikh Abdullah he made two important points:

> I have given careful thought to the suggestion made in the above memo. It appears that there are certain aspects of the question which have not probably struck you. I am, therefore, asking you to reconsider your opinion after giving your earnest consideration to the following:
>
> (1) Firstly in case the administrative control of the State Army is transferred to the Union Government as suggested by you the Pakistan propagandists will make capital out of it and make a plausible case that the State has been completely annexed to Indian Union by taking over complete Military control. Such an impression should not be permitted to be created.
>
> (2) Secondly the Pakistan suggestion that at the time of Plebiscite the Forces of Indian Union should be sent away may to some extent be forced on us. In case the administrative control of the State Forces is handed over it will be difficult for us to make out a case for the retention of the State Forces because administratively there will be no distinction between the two.

However, Sheikh Abdullah was adamant. In a long and angry letter to Jawaharlal Nehru he made a frontal attack upon the State Forces, accused them of all sorts of crimes and misdemeanours, and insisted that their separate entity should cease and that they should be entirely taken over by the Indian Army. To add to the pressure, he informed my father that his Government had decided to stop all the salaries and allowances of these troops with effect from 16 August 1948. The conflict continued for a long time, and it was only some years later that the merger with the Indian Army took place.

Another source of conflict were the 'Reserved Subjects' which were supposed to have been left under my father's authority, including pensions of various types to members of the ruling family, the hospitality (*Tawaza*) department, and so on. Then there was the question of land reforms. The Sheikh and his party were pressing for radical changes in the land tenancy system, while my father was obviously opposed to any such move. The Dharmarth Trust, a religious endowment created by the founder of the State, Maharaja Gulab Singh, which administered over a hundred Hindu temples and shrines in different parts of the State, also became a bone of contention. This Trust spent several lakhs of rupees on the feeding of refugees in Jammu after the invasion, but Sheikh Abdullah charged that Dharmarth money had been used for political purposes, although this allegation was never proved or substantiated, and was anxious that his Government should take over the Trust and its extensive assets. This matter became the subject of an intense and bitter controversy, and ultimately Sardar Patel had to personally intervene and a compromise was worked out whereby my father remained Sole Trustee, but a reputed firm of Chartered Accountants was appointed to undertake a thorough audit of the Trust funds.

Apart from the merits of these and other disputes, the basic fact was that my father and Sheikh Abdullah represented two political cultures so dissimilar and disparate that the possibility of any compromise was virtually non-existent. My father belonged to the feudal order and, with all his intelligence and ability, was not able to accept the new dispensation and swallow the populist policies of Sheikh Abdullah. The Sheikh, on the other hand, while a charismatic mass leader and a superb orator in Kashmiri, was imbued with a bitterly anti-Dogra and anti-monarchical attitude. His socio-economic policy spelt out several years earlier in a pamphlet entitled 'Naya Kashmir' was based on egalitarian and socialistic thinking in which the ruler, at best, would be a powerless figurehead.

Added to the ideological conflict between them was the complicating fact that the Sheikh really represented Kashmiri opinion which, although numerically in a majority, was confined to only one region of the State. My father, in turn, commanded the traditional loyalty of the Dogras who constituted a majority in the Jammu region. Thus the stark contradictions inherent in the multi-regional character of the Jammu and Kashmir State founded by my ancestors surfaced with the events that followed the British withdrawal in 1947. Superimposed upon this whole situation was the fact that the State had become the subject of an international dispute and for many years constituted a permanent item on the agenda of the United Nations, as a result of which the plebiscite offer hung constantly over our heads and constituted an additional element of tension and instability.

The hostility between my father and the Kashmiri leaders was greatly heightened by the fact that his move to Jammu on the night of 27 October 1947 in the wake of the tribal invasion (on the insistent advice of V. P. Menon) was seized upon by the Sheikh to attack and malign him in bitter and brutal fashion. It was presented to the country and the world that here was a cowardly ruler who fled his capital in the dead of night along with his family, jewels and courtiers, leaving his people to face the fury of the approaching onslaught. A barrage of vituperative propaganda was launched by the National Conference leaders, and was echoed by newspapers in Delhi and other parts of the country. The irony of the situation lay in the fact that Sheikh Abdullah himself had flown to Delhi on 25 October, two days before we left Srinagar, and did not return until after the Indian Army arrived. Even worse, my father was publicly accused by Sheikh Abdullah of having fomented communal riots in Jammu in which Muslims had been massacred, and for years he would say in his speeches that the hands of the Dogras were 'dyed in blood'. For a proud and highly sensitive man like my father, these attacks were naturally a source of great anguish and anger. It is to his abiding credit that, despite tremendous provocation, he maintained a dignified silence in public, and for the rest of his life said or did nothing that could have harmed the national interest at that critical juncture.

However, he did protest to Sheikh Abdullah in a memorandum dated 3 December 1948, as follows:

PRIME MINISTER

I would like to draw your attention to the malicious propaganda which is being carried on against my person inside the State and outside. I presume this has come to the notice of the Prime Minister and the Ministry but I find that no steps have been taken either to counteract or ban such activities. I am sending herewith some copies of extracts of the speeches of some of the Ministers and National Conference Leaders which also offend in the same manner.

I am sure you will agree that it is most improper both on constitutional and moral grounds for Ministers to indulge in this kind of propaganda. I put it to you that it should be as much the concern of my Government as of myself to ensure that the person of the Constitutional Head of the State, his dignity and position are fully respected and that any tendency to the contrary, from whatever quarter, is dealt with promptly and severely. I hope you will take suitable action immediately to counter these tendencies and this propaganda. I will be glad if you will let me know as soon as possible what you propose doing.

MAHARAJADHIRAJ
3.12.48

No reply is on file, but Sheikh Abdullah could hardly have been expected to respond favourably. Indeed, he redoubled his demand that my father should be forced to physically leave the State, by abdication or otherwise.

In the conflict of wills with my father, Sheikh Abdullah had a major advantage in the whole-hearted support of Jawaharlal Nehru. As against this my father was closely in touch with Sardar Patel, Minister for Home Affairs and the States Ministry, who often stood up for him against the constant harassment let loose by Sheikh Abdullah. The former Prime Minister of Jammu and Kashmir, Gopalaswami Ayyangar, was also in Jawaharlal's Cabinet, and because of his intimate knowledge of the State, his services were utilized on several occasions in dealing with Kashmir affairs. That this was deeply resented by the Sardar is evident from the correspondence published in the first volume of his letters, which reveals that at one stage this became so serious as to threaten a major rift between the two great leaders, and that the Sardar wrote out (but did not dispatch) his resignation on this very issue.

Ever since the accession, my father maintained close contact with Sardar Patel and wrote to him with regard to a number of important matters. The correspondence includes several appeals on behalf of my father against the constant harassment and humiliation directed at him by the Sheikh, and evidently the Sardar did intervene on several occasions to sort things out. But the main issue at the time was the Security Council deliberations, and the necessity of making drastic changes in the internal administration of the State could not be evaded. It seems that at first the 'Mysore model' was mooted, under which Sheikh Abdullah would be appointed Head of the Administration and my father would have a Dewan to look after the Reserved Subjects. Something on these lines was in fact attempted for a while, but riding a popular wave as he was, Sheikh Abdullah was clearly in no mood to accept anything short of a complete transfer of power. The correspondence also shows that my father was very unhappy with the slow progress of the military campaign against the raiders. This is clear from a long and passionate letter that he wrote to the Sardar on 31 January 1948. After a lengthy description of the unsatisfactory military situation, the never-ending influx of refugees and the deadlocked Security Council deliberations, he made some remarks which I quote at length because they reveal his state of mind at the time, and also contain the first hint of his possible withdrawal from the scene.

In the situation described above a feeling comes to my mind as to the possible steps that I may take to make so far as I am concerned a clean breast of the situation. Sometimes I feel that I should withdraw the Accession that I have made to the Indian Union. The Union only provisionally accepted the Accession and if the Union cannot recover back our territory and is going eventually to agree to the decision of the Security Council which may result in handing us over to Pakistan then there is no point in sticking to the Accession of the State to the Indian Union. For the time being it may be possible to have better terms from Pakistan but that is immaterial because eventually it would mean an end of the Dynasty and end of the Hindus and Sikhs in the State. There is an alternative possible for me and that is to withdraw the Accession and that may kill the reference to the UNO because Indian Union will have no right to continue the proceedings before the Council if the Accession is withdrawn. The result may be a return to the position the State held before the Accession. The difficulty in that situation however will be that the Indian Troops cannot be maintained in the State except as volunteers to help the State. I am prepared to take over command of my own Forces along with the Forces of the Indian Army as volunteers to help the State. I am prepared to lead my Army personally and to command, if the Indian Union agrees, also their troops. It would certainly hearten my people and the Troops. I know my country much better than any of your Generals will know it even during the next several months or years and I am prepared to take the venture boldly rather than merely keep on sitting here doing nothing. It is for you to consider whether the Indian Union will accept this in both the situations whether after the withdrawal of the Accession or even if the Accession continues. I am tired of my present life and it is much better to die fighting than watch helplessly the heart-breaking misery of my people.

So far as the internal political situation is concerned I have left the matter entirely to you personally. I am prepared to be constitutional Ruler of the State and when a new constitution is framed I am quite willing to give Responsible Government but I am not prepared to go beyond the Mysore Model because I am not satisfied that the leaders of the National Conference are for the time being very fit administrators or command the confidence of the Hindus and Sikhs and even of a large section of the Muslims. I must therefore keep certain reserved powers of which you are already aware and I must have a Dewan of my free choice as a member of the Cabinet and possibly as President.

Another alternative that strikes me is that if I can do nothing I should leave the State (short of abdication) and reside outside so that people do not think that I can do anything for them. For their grievance they can hold the Civil Administration responsible or the Indian Forces who are in charge of the Defence of the State. The responsibility will then clearly be either of the Indian Union or of the Administration of Sh. Abdulla. If there is any criticism those responsible can have it and the responsibility for the suffering of the people will not be mine. Of course I well anticipate that—as people started saying when I left Kashmir only on Mr Menon's advice that I had run away from Srinagar—they will say that I have left them in their hour of misery but it is no use remaining in a position where one can do nothing merely to avoid criticism. Of course if I go out of the State I will have to take the public into confidence and tell them the reasons why I am going out.

The Sardar's reply came on 9 February. The operative paragraph was short but categorical. He wrote:

I fully realize what an anxious time you must be having. I can assure you that I am no less anxious about the Kashmir situation and what [is] happening in the U.N.O., but whatever the present situation may be, a counsel of despair is entirely out of place.

Subsequent letters deal with the issue of a proclamation by my father handing over effective power to Sheikh Abdullah, the office and function of the Dewan Th. Baldev Singh Pathania, the definition and administration of the Reserved Subjects, the future of the J & K State Forces and, of course, Sheikh Abdullah's pronouncements. In a revealing letter to the Sardar dated 20 April 1948 my father wrote:

My dear Sardar Patel,

As I have mentioned to you and to Mr Menon and Mr Shankar also once or twice, there is one aspect of propaganda against me which has distressed me beyond words, particularly as it affects not only my position as a Ruler but my personal honour. I refer to the wild and baseless allegations that are being made against me that I left the Capital at the dead of night and removed truckloads of furniture and other belongings. I would ordinarily have dismissed these allegations but to my great regret and profound shock, they have found expression in some of the utterances of my present Prime Minister. I would invite your attention to the speech which Sh. Abdulla made before the Security Council in which he said the following words:

> 'The Maharaja in the dead of night left the Capital along with his courtiers and the result was absolute panic. There was no one to take over the control. This is the manner in which administration changed hands and we were de facto in charge of the administration. The Maharaja later on gave it a legal form.'

(2) How entirely divorced from facts this allegation is would, I am sure, be borne out by your Secretary Mr Menon at whose instance I reluctantly left Srinagar for Jammu. The other allegation that truckloads of belongings were removed is absolutely false and fantastic. The fact is that some lorries carrying the families of those officers who had already moved to Jammu on Government work and families of servants followed me. I felt morally responsible to afford transport facilities to these families. All my household effects which usually remain permanently at Srinagar have been and are still there. I am sure you will appreciate that such speeches and allegations when made by responsible and highly placed persons in the administration and by members of the National Conference are bound to create disaffection and promote feelings of estrangement between the Ruler and his people.

(3) You might ask why I am resuscitating this matter after so much lapse of time since the speech was made, but apart from the fact that those allegations are still being persisted in and are being given credence under continuous propaganda, the speech of my Prime Minister is being distributed in a pamphlet form headed 'Kashmir: Appeal to World Conscience'. I hope your Ministry will succeed in giving a lie to this propaganda by means of an authoritative pronouncement or communique explaining the correct facts and by persuading Sh. Abdulla to make amends for the lapse which he has made.

(4) While I am on this subject I should also like to refer to one important aspect of the propaganda which is being carried on against me on the lines that I was a despotic and autocratic ruler, that the popular

movement was directed against Dogra tyranny and that the present position has been reached as a result of fight put up by the people against me and my regime. Apart from the fact that I am prepared to maintain and establish to the satisfaction of any impartial person that at any given time, since 1934, the reforms which I gave to my people were in advance of those given by the Ruler of any State in India and that my attitude in relation to India's constitutional advancement has been much more progressive than that of any other Ruler I should like to ask whether any useful purpose could be served by raking up the past. For my part, I am endeavouring to forget all that was unpleasant in the past. Am I wrong in claiming a similar response from the persons in power in relation to me? Moreover, by presenting me as an unmitigated autocrat it is perhaps not realized that they merely help Pakistan propaganda. I should have thought that they would themselves realise the expediency of countering this propaganda of Pakistan rather than keep afresh old animosities, for, as some knowledgeable persons have told me, the justification of Poonch and Mirpur revolt can be sought in the pages of some of the publicity material issued by the Ministry.

(5) I hope that instructions from you to Sh. Abdulla on this point would go at least some way in repairing the damage done. I can assure you that had I not felt both these matters to have an important bearing from the propaganda point of view in order to win the State of Jammu & Kashmir for a permanent association with the Indian Dominion, I would not have laid so much stress on them.

Yours sincerely,
Hari Singh

It is clear, however, that even the Sardar was not in a position to restrain Sheikh Abdullah, and by the time I returned to India in February 1949 the situation had become quite impossible and a total deadlock reached. In March I turned eighteen, but the event passed off virtually unnoticed as the atmosphere was full of tension and foreboding. Soon thereafter my father received an invitation from Sardar Patel suggesting that he, my mother and I should visit Delhi for consultations. We left in April in a chartered DC-3. Little did I realize as we boarded the plane that only my father's ashes would ever return to his beloved Jammu.

Chapter 8

On our arrival in Delhi we stayed first at the Maidens Hotel in old Delhi and later shifted to the Imperial. Soon after we got there my father, my mother and I were invited by Jawaharlal Nehru to lunch with him at Teen Murti House. Indira Gandhi was hostess, and though my father and Jawaharlal were obviously uncomfortable in each other's presence, there was no open hostility. Evidently Jawaharlal had left it to the Sardar to get down to brass tacks with him. On 29 April we had a meal with the Sardar also, at which his daughter Maniben and his Private Secretary V. Shankar were present. After dinner my parents and the Sardar went into another room, and it was there that the blow fell. The Sardar told my father gently but firmly that although Sheikh Abdullah was pressing for his abdication, the Government of India felt that it would be sufficient if he and my mother absented themselves from the State for a few months. This, he said, was in the national interest in view of the complications arising from the plebiscite proposal then being actively pursued in the United Nations. He added that as I had now returned from America, I should be appointed Regent by my father to carry out his duties and responsibilities during his absence.

My father was stunned. Although rumours to the effect that he might be pushed out of the State had been in the air for some time, he never believed that even the Sardar would advise him to adopt this course. He emerged from the meeting ashen-faced, while my mother was fighting desperately to keep back her tears. The Sardar also looked grim as he saw us to the door. I realized at once that something was seriously wrong, but didn't dare ask what the matter was. We drove back to the hotel in dead silence. My father immediately got into a huddle with his advisers, Bakshi Tek Chand and Mehr Chand Mahajan, and his staff officers. My mother went to her room where she flung herself on her bed and burst into tears. I followed her there, and when she had calmed down a bit she told me that she and my father were being pushed out of the State and that the Government of India wanted me to be appointed Regent.

It was at this critical juncture that Jawaharlal invited me to the first of several breakfasts at his residence. I can still recall my excitement as I walked up the steps of Teen Murti House to the dining-room. The table was set for two, Indira Gandhi and the boys having already finished their meal. Jawaharlal walked in briskly, shook hands with a friendly 'Hello, Tiger', and we sat down to breakfast. Over the meal, which lasted almost an hour, Jawaharlal asked me a few general questions and then settled down into a long monologue. In his chaste accent he spoke of the fact that a new India was being created, that the old feudal order was rapidly collapsing and that as a young man I should readily adjust myself to the new situation. He then went on to outline the Kashmir problem, the role of Sheikh Abdullah and the importance in the national interest of establishing a harmonious situation in the State. He said briefly that my father was evidently unable or unwilling to accept the new dispensation, and that both he and Sheikh Abdullah were of the view that I should be appointed Regent so as to solve the present deadlock. When he spoke of the future of India his eyes glowed with a gentle fire and his voice took on a ringing note.

I had admired this man ever since I was a schoolboy, and it was a fascinating experience to be near him and to listen to him. When he talked of the vast, historic forces that were sweeping across the world it was as if the *Discovery of India* was coming alive before my eyes. How different this was from the constricted viewpoints that I heard in my father's circle and the all-pervasive atmosphere of tension and intrigue at home. In my school days I had dreamed of becoming a perfect ruler, of doing something substantial for the people. I now realized that rulership had gone, never to return, but here was an even more exciting opportunity of doing something in the broader national interest, and at the personal behest of one of the greatest leaders of our times.

For the next few days the tension steadily mounted. We had moved into the Imperial Hotel where we occupied several suites, and there was a constant stream of people coming in to see my father. Prominent among them was V. Shankar who, with the Sardar's health steadily deteriorating, was becoming increasingly important in the States Ministry. Sardar Patel called me over one evening and had a long, friendly talk about the whole situation. While Jawaharlal's conversation was in the broader context of Indian nationalism, the Sardar chose a narrower canvas. He said that while he fully realized and admitted that an injustice was being done to my father, he had to agree to this in the broader national interest mainly due to Sheikh Abdullah's insistence. He also said that I should face the situation bravely and undertake my new responsibilities with courage and confidence. His talk was a source of great encouragement to me at that difficult juncture. Soon thereafter the Sardar left for Dehra Dun, and it was decided that my parents and I would also go there later to visit him. Meanwhile, my father was grappling with the virtual ultimatum served upon him by the Government of India, and was gradually coming round to accepting the inevitability of leaving the State. There was some pressure upon me to decline the Regentship, but I indicated as gently as I could that I did not think it would be appropriate for the whole family to leave the State, thus severing all ties with Jammu and Kashmir.

I had barely turned eighteen, but due perhaps to the unusual circumstances in which I had grown up I discovered that I was quite capable of holding my own in discussions even with very senior leaders. While I showed no outer signs of lack of self-confidence, inwardly, I must admit, I was often on the verge of panic. The last few years had been so full of change, and my own prolonged absence in America had also contributed towards creating a strange and unsettling atmosphere. All the familiar landmarks of my childhood had disappeared, and even my mother, with whom I had such close emotional links, seemed to be on the verge of psychological collapse. In fact, her position was excruciatingly difficult; on the one hand was her loyalty to my father despite their temperamental differences, and on the other was her devotion to her only child. Characteristically, my father never discussed the problem with me directly. The dialogue would take place either through his Private Secretary, Pandit Bhimsen Mahey, or Mehr Chand Mahajan, who remained his close confidant even after relinquishing the Prime Ministership. It seems that after the initial shock my father reconciled himself to leaving the State for a while, although I think he realized that it would not be easy for him ever to return. He was, however, totally opposed to abdication, and reiterated this in a letter written to Sardar Patel on 6 May. I reproduce below the full text of his letter and of the reply that the Sardar sent from Dehra Dun, because they have a vital bearing on subsequent developments in my life.

My dear Sardar Patelji,

With reference to the discussions I had with you on 29th April and 1st May 1949 I have been revolving the matter in my mind and am now in a position to let you have my settled reactions to the proposal in regard to my temporary absence from the State which you put to me.

I should like to say at the outset that I was completely taken aback by this proposal but coming as it did from you, in whom I have since the very beginning placed implicit trust and confidence and whose advice I have throughout followed on the many questions affecting me personally and my State both in the present and in future, I have been able somehow to adjust myself to it. I would not, however, be human if I did not express my sense of keen disappointment and bewilderment at having been called upon to make such a sacrifice of prestige, honour and position when all along I have been content to follow, sometimes even against my own judgment and conscience, the advice in regard to the constitutional position in the State which I have been receiving from the Prime Minister of India or yourself, sometimes even against arrangements which were agreed to only a few months before. Nor would it be fair on my part to conceal from you my own feeling that while Sheikh Abdullah has been allowed to depart from time to time as suited his inclinations from the pledged and written word, to act consistently in breach of the loyalty which he professed to me prior to his release from Jail and the oath of allegiance which he took when he assumed office, and to indulge openly along with his colleagues in a campaign of vilification and foul calumny against me, both inside the State and outside, I should have had to be driven from position to position—each of which I thought I held on the advice of the States Ministry.

The contrast naturally fills me with poignant feelings. However, once again putting my complete trust in your judgment and benevolent intentions towards us, I might be prepared to fall in with your wishes and to absent myself from the State for a period of three or four months in consideration of the fact as emphasized by you namely complications created by the reference to U.N.O. and the plebiscite issue.

There are, however, certain questions arising out of this proposal on which I would venture to make my position clear to you and on which

I would be grateful to have your assurance. I hope you will kindly appreciate the necessity of my seeking these assurances. I have to think of the immediate future in the light of my bitter experience of the last several months and I owe it to myself, my family and my dynasty to procure a clear declaration in respect of these matters.

1. I should like to be assured that this step is not a prelude to any idea of abdication. I should like to make it clear now that I cannot entertain the latter idea even for a moment and am fully prepared to take the consequences. I regard such a demand from my Prime Minister and his colleagues as a clear breach of the many understandings on which constitutional arrangements have been based from time to time and a positive net of his disloyalty, treachery and deception.

2. Sheikh Abdullah should be clearly told to stop the campaign of vilification against me and to abandon all activities, both on his part and that of his followers, aimed at securing my abdication. I feel that the sacrifice which I am being called upon to make would be in vain if I continued to be the target of their public or private attacks.

3. There should be a clear assurance of protection of myself and my adherents against any victimization. In this connection I should like especially to draw your attention to the facts that have been reported to me about persons having been detained in jail for their failure to sign for my abdication.

4. The question that I should remain out of the State for three or four months for reasons of health, will, I am afraid, not be believed by anybody and is likely to give rise to many misgivings and speculations within and outside the State as

(*i*) everybody knows that I am not in such a state of health as would necessitate a long rest outside the State. I have, on your advice, been recently touring parts of the Jammu Province in the heat of April.

(*ii*) For everybody in bad health Kashmir is considered to be the best health resort and it will certainly look strange if I went outside the State giving out that I am doing so for reasons of health.

(*iii*) Wherever I take my temporary residence I cannot confine myself to the four walls of the house. I am bound to meet people, who, when they meet me, will never believe that I am staying there for reasons of health.

(*iv*) Some other reason which may be plausible and may also at the same time not compromise my dignity and position, should be given out. The best thing would be that the Government of India should find a suitable position for me in Delhi where my services may be utilized in a fitting manner during the above period of 3 or 4 months.

5. It is a matter of paramount necessity that Her Highness should remain with the Yuvaraj in the State during the period of my absence.

He is young and impressionable and requires paternal guidance and personal supervision of at least one of the parents. I can see no reason either of political expediency or justice in insisting on the separation of a mother from her only child whom she is seeing after thirteen months of absence abroad. Considerations of humanity alone should suffice to rule out this altogether.

6. My Private Estates, Houses and other property should be protected against the aggressive acts of Sheikh Abdullah's party. They will attempt to take possession of my Houses, Gardens, Lands and other property. The Indian Dominion should guarantee against that act of aggression. While I am there they dare not do these things but in my absence they will attempt this. I have received information that even during the last few days, after I left Jammu for Delhi, encroachments have been made on my lands at Srinagar.

7. No change should be made without my consent in the present arrangements regarding the State Forces or the constitutional position, Prerogatives etc. of the Ruler as now subsisting. Arrangements will continue for me to draw my Staff (both State and Private Deptts) from amongst Officers of my Forces. Guards mounted by my Forces at my Palaces will also continue as at present as per agreement reached vide my letter of 30th August and Mr Menon's reply of 3rd September thereto. I shall also take whatever staff etc. I require with me outside.

8. I should be entitled during my stay in India to suitable strength of military guards wherever I stay.

9. Yuvaraj's safety and protection should be the concern of the Indian Dominion. State and Indian Military should guard his person.

10. Outstanding matters with the State Military, Civil Lists, Hazur Departments etc. should be decided with me immediately.

In conclusion I wish to say that I shall take the final decision on getting assurances from you on the points above mentioned.

With kindest regards,

Yours very sincerely,
Hari Singh

Imperial Hotel,
New Delhi.
6th May, 1949.

After about a fortnight, Sardar Patel's reply came from Dehra Dun.

Camp: Doon Court,
Dehra Dun.
23rd May 1949

My dear Maharaja Sahib,

Thank you for your letter dated the 6th May 1949.

2. I am very glad to know that Your Highness has reconciled

yourself to the proposal which I put forward at my discussion with you. It was with no light heart that I did so. No one can be more cognizant than myself of the attitude which Your Highness has adopted ever since you signed the Instrument of Accession. I am grateful to Your Highness for the spirit of co-operation and understanding which you have always extended to me and also for the kind sentiments which you have expressed. I can assure Your Highness that, before putting forward my proposal, I had, after careful consideration, come to the conclusion that the interests alike of Your Highness, the dynasty and the country demanded the step which you have now agreed to take. I know full well the personal sacrifice involved in it, but, I am sure, along with so many other changes to which Your Highness has accustomed yourself, you will undertake this step also with a sense of duty to your country and in a spirit of calm resignation to the superior dictates of events.

3. Regarding the points which Your Highness has referred to me, I should like to state that the question of Your Highness' abdication does not arise. We have made the position quite plain to Sheikh Mohammad Abdulla, and we hope there will be an end to the public controversies centering round this matter as well as to the derogatory references to Your Highness in the press and on the platform in the State. Your Highness will, of course, appreciate that the future constitution of the State would be determined by the duly elected constituent assembly. I am afraid, in the absence of any specific instances of victimization to which Your Highness refers in paragraph 4, it may not be possible for me to give any assurance, but I can tell Your Highness that, if any such instances are brought to our notice, we shall look into them and try to see that justice is done.

4. I appreciate what Your Highness says in regard to the reasons for your remaining outside the State, but I feel that it would be best just to say that Your Highness has decided, after the strain of the last so many months and continued ill-health, to stay out of the State for a few months. The actual period need not be stated.

5. We have carefully considered the question of Her Highness staying with the Yuvaraja during your absence, but for a variety of reasons, we feel that it would be best, for the present, for her also to stay away for a while. Later, she can certainly visit the Yuvaraja from time to time, and the Yuvaraja can also visit Your Highness and Her Highness occasionally.

6. We would be grateful if Your Highness would let me have a list of the private estates, houses and other property belonging to Your Highness and referred to in para 6 of your letter. On receipt of the list we shall take up the matter with your Ministry. In the meantime, I hope that, with the understanding that has been reached with Sheikh

Mohammad Abdulla on the various controversial issues, he will himself take steps to safeguard Your Highness' property against encroachments. In particular, I hope that the feelings which he entertains for the Yuvaraja will succeed in closing the chapter of the past several months and in introducing a healthy change in the approach of both the Government and the National Conference workers to these and other problems affecting Your Highness and the family personally and the dynasty in general. I hope that no need for a change in the arrangements to which Your Highness has referred in paragraph 7 would arise, but should any such need arise, we would of course consult Your Highness. We would also make necessary arrangements for guarding you during your stay in India, and we take full responsibility for the Yuvaraja's safety and protection.

7. As regards outstanding matters, we have already informed Your Highness separately that your Civil List has been fixed at Rs. 15 lakhs, out of which Rs. 6 lakhs would be paid by the State and Rs. 9 lakhs by the Government of India during the period of the emergency. Out of this sum, Your Highness will have to make allotment to Her Highness and the Yuvaraja. The latter's expenses would, of course, be larger than before on account of his being Regent. I hope Your Highness will agree to make a suitable allowance for him, bearing in mind his needs. I should be glad to know your suggestion in this respect. Similarly, I should welcome Your Highness' proposal regarding Her Highness' allowance. As regards Hazur Departments, Sheikh Sahib has already agreed to our suggestion that an allowance of Rs. 5 lakhs should be made to Your Highness for this purpose. A list of items, which will be controlled by Your Highness, and during your absence, by the Yuvaraja as Regent, has been furnished to Sheikh Sahib, and he has promised to let us have his comments as soon as possible. After his comments are received, we shall finalise the whole matter, but in the meantime, Your Highness can make suitable allocation to major and other heads of the expenditure on State Departments from out of the allotment of Rs. 5 lakhs.

With kindest regards,

Yours sincerely,
Vallabhbhai Patel

The same day, Sardar Patel sent a copy of this letter to Jawaharlal Nehru. The covering note contains the following paragraph: 'As regards the Yuvaraj, we have had a very detailed talk with him and I have impressed upon him the significance and importance of the agreements reached and the consequences which flow therefrom. He is a sensible lad and I think he appreciated the situation fairly well and realizes his responsibilities. He is, of course, still in his teens and would require some guidance. I am looking out for a suitable adviser for him on whose advice he can lean. We shall have to be very careful in the choice of a suitable person.'

Soon after this correspondence took place, all of us went up to Dehra Dun at the suggestion of Sardar Patel who was himself convalescing there. My parents stayed in a hotel, but at the Sardar's special invitation I was his guest for three weeks in the famous Circuit House known as Doon Court, set in an extensive estate with lovely flowers, trees and shrubs. Evidently he and Jawaharlal had decided that it would be useful for me to stay with him for a while before taking up my new responsibilities. The Sardar was at that time in very poor health, and his daughter Maniben was constantly ministering to him with great devotion. He ate mainly in his own room, but would sometimes call me and talk about Kashmir. While he did not have the fire and fervour of the Prime Minister, he spoke with a quiet confidence that was very impressive. Here was a man, one felt, for whom no problem was too formidable to tackle. It was only about Kashmir, which Jawaharlal was handling himself, that the Sardar was obviously unhappy. Although he never criticized Jawaharlal in my presence, it was clear from his conversation that he did not approve of his special relationship with Sheikh Abdullah, whom he evidently distrusted and disliked.

While at Dehra Dun I was able to revisit the Doon School for the first time since leaving it three years earlier. By now I was walking without the aid of a cane, although I would carry a limp for the rest of my life. It was curious coming back to school and seeing again the rooms and grounds where I had spent so many years as a boy. Mr Foot had returned to England and J. A. K. Martyn was Headmaster. Jack Gibson, still Housemaster of Kashmir House, greeted me like a long lost friend and we played several games of chess at his cottage. I also went up to Mussoorie for a weekend which we spent largely with Lakshmi Nivas Birla and his family who were close to the Sardar and V. Shankar.

My parents returned to Delhi earlier, and by the time I got back it had been decided that my father would appoint me Regent and that I would fly up to Srinagar on 20 June to take up my new responsibilities. I learnt that my mother had also agreed, very reluctantly, to leave the State but that, as she could not take the heat of Bombay, she would go to Kasauli instead. One thing remained to be done before I went up to Srinagar. A devout Hindu, my mother insisted that my sacred thread ceremony should be performed. This was done at 5 Hastings Road, where Mehr Chand Mahajan lived. The Pandits came down from Jammu, and two young Rajput servants of my father went through the ceremony along with me. We all got into dhoties, but, despite the Raj Pandit's insistence, my voluble protests saved my head from being shaved as is traditionally done on this occasion. Instead, a symbolic tuft of hair was cut off, saving me from the fate of having to assume the Regency of Jammu and Kashmir looking like a monk.

On the night of 19 June I could hardly sleep. My mind was a welter of conflicting emotions and thoughts. I was obviously embarking on a delicate assignment fraught with risk and even danger. My father made no secret of his unhappiness at the whole situation, and although he had bowed to the inevitable I realized that our relationship had been soured in the process. My mother also was breaking away on her own, which added a new element of emotional instability to the entire situation. I had no one to rely on except Jawaharlal's support and such inner resources as I could summon at that critical juncture. 20 June 1949 was an important landmark in my life. My father, his staff and servants left by train early in the morning for Bombay. My mother, uncle and her maidservants left by car for Kasauli soon after. For half an hour I sat absolutely alone in my hotel room, poised between the weight of the past and the burdens of the future. It was one of those critical moments in life that remains for ever embedded in one's memory. With a considerable effort I pulled myself together and walked out of the hotel accompanied by my staff officer, Captain Mahel Singh. We drove to Safdarjung airport where a DC-3 was waiting to fly me to Srinagar along with Vishnu Sahay, Secretary for Kashmir Affairs, a senior I.C.S. officer who had been deputed by the Prime Minister to accompany me.

The proclamation that my father had signed on that very morning, before he left, was short and cryptic. It read as follows:

PROCLAMATION

Whereas I have decided for reasons of health to leave the State for a temporary period and to entrust to the Yuvaraj Shri Karansinghji Bahadur for that period all my powers and functions in regard to the Government of the State.

Now, therefore, I hereby direct and declare that all powers and functions, whether legislative, executive, or judicial which are exercisable by me in relation to the State and its Government, including in particular my right and prerogative of making laws, of issuing Proclamations, Orders and of pardoning offenders, shall during the period of my absence from the State be exercisable by the Yuvaraj Shri Karansinghji Bahadur.

Hari Singh
MAHARAJADHIRAJ

As the plane took off it struck me that two critical events in my life had been associated with an air flight. I remembered flying off to America as an invalid, and then returning to the curiously muted welcome in India. Once again I was flying off, almost literally into the unknown. Of course, the valley was well known to me, but I realized that the Kashmir I was now going to was substantially different from the one I had known thus far. Although I was going as Head of State, it was virtually on sufferance of Sheikh Abdullah who wielded effective power. In one of his petulant moods my father had said over lunch, much to my mother's irritation, that even if I went as Regent I would be unceremoniously thrown out in disgrace by Sheikh Abdullah within a few months. Though I shared my mother's resentment at this grim prophecy, I could not easily dismiss it when I considered the political developments over the last few years.

We were scheduled to fly direct to Srinagar, but due to bad weather and the primitive aviation aids at that time, we had to land in Jammu and stay there for an hour before proceeding. When the plane took off from Jammu we were engulfed in thick cloud. It was only as we crossed the Banihal Pass that the clouds broke and the valley of Kashmir lay before us, glittering in all its indescribable beauty like some rare jewel nestling in the heart of the mighty Himalaya.

Chapter 9

At the Srinagar airport Sheikh Abdullah and his entire Cabinet, along with senior officials, had gathered to greet me. The Sheikh met me at the top of the steps and then introduced me to his colleagues. At that time there were nine Cabinet Ministers, important among whom were the Deputy Prime Minister, Bakshi Ghulam Mohammad, the main organizer in the Sheikh's party, and Mirza Mohammed Afzal Beg, the wily Revenue Minister who had been in the Government even under my father during a short-lived experiment in self-rule a few years earlier. Ghulam Mohammed Sadiq, Girdhari Lal Dogra, Sham Lal Saraf, Colonel Pir Mohammad and Sardar Budh Singh were also in the Cabinet, while the bearded Maulana Masoodi was General Secretary of the National Conference.

After the welcome we all drove up to Karan Mahal. It was exciting coming back after so many years in vastly different circumstances to the house where I had spent much of my childhood. I still remember the mingled feelings of exhilaration and apprehension as I alighted from the car and entered the lovely villa, its garden ablaze with flowers and a cool breeze coming in from across the lake. It had been miserably hot in Delhi, and merely getting away from the heat was a great relief. Vishnu Sahay, who was functioning as part-time Adviser, stayed in the guest cottage across the road, and spent a fortnight in Srinagar before returning to Delhi. He was an able man, unflappable and shrewd in his observations, and in the early days of my Regency was a help to me in putting things in their proper perspective without any emotionalism or fanfare.

Although I savoured my newly acquired independence, I fully realized that I was actually in a rather difficult position. Sheikh Abdullah was the outstanding personality on the scene, and Jawaharlal Nehru had specifically asked me to work in harmony with him. Physically an impressive man, the Sheikh was at that time at the height of his powers. He would come over from time to time and give me long lectures on the political situation. He never mentioned my father directly, and I found that he displayed some genuine friendliness towards me. I was aware, though, that my position was a precarious one and that in a way I was on a trial run. My father was not really happy at my having become Regent, and there was a certain sullen disapproval from the Dogra public in Jammu. In all my actions, therefore, I had to try and steer a middle path between appearing subservient to the Sheikh on the one hand and offending him and Jawaharlal on the other.

Meanwhile there was a lot of interesting work to be done. The Kashmir issue being very much in the fore at the United Nations, a great deal of significance was attached to the U.N. observers and representatives who were in Srinagar at that time, headed by a Frenchman, General Delvoi. The group was drawn from various European and South American countries, and several had brought their families with them. There were parties at various places, including the residence of the dashing and witty Divisional Commander, General T. S. Thimaiyya and his wife Nina. 'Timmy' was great fun, and exuded charm and charisma. He was adored by his troops and was also very popular with the civilian administration. Soon after I got up to Srinagar I went with him to some of the forward areas on the Ceasefire Line. We crossed the 10,000 foot Nashtachun Pass and visited Tangdhar and Tithwal. It was my first exposure to the greater Himalayas beyond the Kashmir valley, and an unforgettable experience. Looking at the mountains from my Srinagar house was one thing; being in them was quite a different sensation. The air was crisp, and the wind swished continuously through the great deodar trees which covered large areas.

Timmy was in high spirits, driving our jeep himself and regaling us with graphic details of the battles fought to free these areas from the invaders only a year ago. The troops were in fine fettle and seemed genuinely pleased to see us. It was the first of numerous encounters I have had with Indian Army personnel living under the most difficult conditions, and I have never failed to be impressed by their devotion to duty and their uncanny capacity to adapt to the environment. The cordiality between the army and the civilian population was also impressive. Once across the Pass, the people were Panjabi- not Kashmiri-speaking, but despite their close physical and ethnic proximity to Pakistan-occupied territory I saw no trace of hostility or resentment. In fact, I was probably the first member of the Dogra ruling family to visit those areas, and they greeted me with great affection. Later we did a similar trip across the Zojila Pass where, in a historic manoeuvre often compared to Hannibal's crossing of the Alps with his elephants, Timmy had taken tanks for the first time in human history to such heights and had routed the startled invaders. On the other side of Zojila we stayed for a night at the small village of Dras, believed to be one of the coldest inhabited places in the world.

In addition to others, I met at this stage a person who was to play a significant role in my political life, Durga Prasad Dhar. Handsome, smooth and brilliant, 'D.P.' was the scion of one of the most distinguished Kashmiri Pandit families in the valley, but very early he had become actively associated with the National Conference. Although only a Deputy Minister, he was widely accepted as being the real brain behind Sheikh Abdullah's regime. He was excellent at army–civil liaison, and acted as a valuable link between the State Government and the Government of India. He had a remarkable mind which, however, he turned generally to brilliant manoeuvring rather than solid achievement. He was younger than the other Ministers, and we got on well together. This friendship was to pay useful dividends in the crisis that lay only four years ahead.

Jawaharlal made two visits to Srinagar during the summer of 1949, and I went to Delhi twice and stayed with him at Teen Murti House. He was a charming host, and despite his crushing schedule always found time to inquire after his guests. We would usually meet over meals, and it was here that I first met Pandit Govind Ballabh Pant, a large man with a walrus moustache and a shaking head that belied his sharp and shrewd mind, and the wiry, mephistophelian Krishna Menon. Padmaja Naidu was also living in the house, exuding good humour and *joie de vivre* despite her constant ill health. Many years later after Jawaharlal had passed away, she became a close friend and we discovered that she was an unusually warm and affectionate human being. As far as Jawaharlal was concerned, his physical and mental agility always impressed me. When he entered a room it was as though a breath of energy had blown in. He moved with a springy gait, as if ready at any moment to break into a run. And yet when he settled back on a sofa to talk his face would go thoughtful, and a strange sadness would cover his handsome features. He spoke beautifully, elocuting each word in a soft, civilized accent. But it was his smile that was the most memorable—pensive and strangely appealing.

During that summer in the valley I revisited many of my childhood haunts, specially Dachigam and Thricker. I did some shooting and fishing, and hosted endless lunch and tea parties for visiting dignitaries and local luminaries. I had just turned eighteen but was full of confidence and subdued excitement at being on my own for the first time. I visited a number of the famous Hindu shrines in the valley, which my father seldom, if ever, went to. His approach to religion had been largely formal, but taking after my mother I regularly performed some puja every morning before going into circulation, and also developed an interest in temples and shrines.

One incident occurred which preshadowed the tensions that lay ahead. The annual festival at the famous shrine dedicated to the Goddess Khir Bhawani fell on a certain auspicious day in June. The Kashmiri Pandits led by Pandit Parmanand, an able officer of great integrity who had just retired as Accountant General, were keen that I should visit the shrine on that occasion. It was clear that despite Sheikh Abdullah's secular protestations the tiny Pandit minority was not feeling very happy under the new dispensation. I readily agreed to go, and my visit was announced. Soon thereafter I got a message from Sheikh Abdullah saying that it was better if I did not go on that day. This placed me in an awkward dilemma, but having given them a commitment I felt I could not let the Pandits down. I received a tumultuous reception at the shrine where thousands had assembled. Elderly ladies embraced me and kissed me on the forehead, a gesture reserved only for those very dear to them. It took me almost three hours to complete the puja around the sacred pool, in which the water mysteriously changes from time to time. It is believed that the colour presages future events, and that dark colours are an ill omen. It is a curious fact that for months before the Pakistani invasion the water had turned pitch black, and also that, though the shrine remained for many months in the area under the occupation of the invaders who destroyed all other Hindu, Sikh and Christian places of worship, not a leaf in Khir Bhawani was touched. Throughout my presence at the shrine the air was rent with greeting slogans, including some of *Dogra Raj Zindabad*. Although the Sheikh never mentioned the matter to me, I learnt that he was very upset that I went there. How could the Lion of Kashmir tolerate any rival, even though he was yet only a tiger cub?

To Pandit Parmanand also I owe a deep debt of gratitude for having introduced me to Sanskrit. I had never studied a word of that magnificent language either at school or college. It was Pandit Parmanand who urged and insisted that I should begin learning Sanskrit, and he would come over to my house thrice a week to teach me some simple verses. Thus it was that, as in the case of English, I learnt Sanskrit not through grammar but directly by ear, and soon found that I could remember and recite passages by heart without undue difficulty. The key lay in the superb musical and metrical quality of Sanskrit verse, whether it is the liturgical majesty of the Vedas, the glowing wisdom of the Upanishads, the immortal teachings of the Bhagavad Gita or the superb hymns of Adi Shankaracharya.

Another development at this point had an important impact on my future life. Soon after I got up to Srinagar I thought it might be a good idea to renew my contacts with the academic world. I asked the Director of Education to suggest the name of a scholar who could visit me regularly to discuss current economic and political problems. He sent a Professor P. N. Chaku, with whom I immediately struck up a valuable rapport. Professor Chaku was an accomplished intellectual and introduced me to a number of important political and economic concepts, including the ideas of Keynes in the field of economics, Laski in political science and Marx in the context of a developing economy. I discovered again a genuine interest in the world of ideas, and began reading extensively. At that time Bertrand Russell and Aldous Huxley were the two authors who influenced me most. Russell's translucent prose and Huxley's imaginative formulations had a deep impact upon me. I still have a letter each from these great writers in response to my 'fan' letters, and despite disagreement in some matters my respect for them remains intact.

In my capacity as Regent I replaced my father as titular

Commander-in-Chief of the Jammu and Kashmir State Forces. By that time these troops had for all practical purposes become part of the Indian Army, and yet unlike the numerous other forces of Indian States that had acceded to the Indian Union, ours had not been formally merged and still retained their separate entity. In effect, of course, the office of Commander-in-Chief was a purely formal one and involved only signing a few papers from time to time. Also formal, but closer to my own interests, was the office of Chancellor of the newly formed Jammu and Kashmir University that I assumed ex-officio as Head of the State. Before 1947 educational institutions in the State had been affiliated to Panjab University based in Lahore, but after Partition the need for a separate university in the State was urgently felt. This was set up through an Ordinance issued by my father in 1948, as there was no legislature in the State at that time. The first formal convocation of the new University was held in Srinagar on 24 September 1949 over which, as Chancellor, I presided. Jawaharlal came up to Srinagar for the occasion. I had sent him an advance copy of my speech and in a letter of 11 September he wrote to me: 'Sheikh Saheb arrived here this evening and gave me a copy of the speech you intend delivering as Chancellor of the University. I have just glanced through it and I find it very interesting. Surely you must be by far the youngest Chancellor that a University has had.'

The convocation itself was a colourful affair. We all wore black gowns and pink turbans, and entered the enclosure in solemn procession. Sheikh Abdullah was Pro-Chancellor, so he, Jawaharlal and I walked together. After giving away degrees and prizes, I read out my speech and then requested Jawaharlal to address the gathering. I started then on a principle that I have consistently followed of writing my own speeches. The only exceptions, of course, were the Addresses to the Legislature which I had to deliver every year, but even for those I would ask for a draft and then try and put the material into my own words. Evidently my speech that day was well received, as I was widely congratulated by all present after the event, but perhaps this was only because of the novelty of having a teenage Chancellor.

In the afternoon we were all taken out in a boat procession on the Jhelum. I remember having sat with my father on one of these processions many years earlier, where brocade covered barges were rowed upstream by dozens of well-trained Kashmiri boatmen wearing white uniforms and yellow turbans. The boats and the boatmen were still there, but now a decade later the whole context had drastically changed. Jawaharlal, Sheikh Abdullah and I sat in the main barge, while other members of Nehru's party, including Gopalaswami Ayyangar, Sardar Baldev Singh, Rajkumari Amrit Kaur and N. V. Gadgil (all members of his Cabinet), as well as the Commander-in-Chief of the Indian Army General Cariappa followed in smaller boats. The banks of the Jhelum were crowded with the people of Srinagar who turned out in tens of thousands to greet Jawaharlal. As we passed under each of the historic seven bridges it seemed as if they would collapse under the sheer weight of people waving greeting flags and chanting slogans. School children in their thousands, smartly turned out in their uniforms, added to the life and colour of the whole scene. It was a dramatic and memorable event for me, and I felt proud to be with a national leader of Jawaharlal's stature under those circumstances. It was also an impressive show of political strength on the part of Sheikh Abdullah and his National Conference, and belied the vicious propaganda relentlessly indulged in by Pakistan to the effect that the Kashmiris were groaning under the military occupation of Indian troops. Anyone who witnessed that procession—and there were a number of foreign newsmen and camera teams present—could not have failed to be impressed by the genuine outpouring of love, affection and hope that Jawaharlal evoked at that historic juncture.

The Kashmir season was now drawing to a close, and we began making plans to go down to Jammu for the annual 'darbar move'. Then, as now, the State Government was based for the six summer months (May through October) in the summer capital, Srinagar, and for the other six months in the winter capital, Jammu. This is a political imperative, because in the State of Jammu *and* Kashmir both major regions have to be given equal importance. Indeed, it was in the inability to hold the scales even between the two regions that the seeds of much future tension and conflict are to be found. I decided that we would move down on 7 November, but before that I thought

I would get in some more fishing and shooting which improves as the autumn advances towards winter. I had even begun playing some tennis despite my stiff hip. My plan was to stay in Jammu for a few weeks and then call on my father in Bombay. He had heard from the Nepal Ranas, and my wedding had been fixed for 15 January next year. Everything seemed to be getting along smoothly.

On 29 October a Chakor shoot had been arranged at Khrew, a few miles out of Srinagar. We set off in the morning; I sat in the front seat along with the driver, while Brigadier Rawat, Captain Mahel Singh, my ADC, and my black labrador retriever, Dusky, got in behind. It had rained the previous night, and except for the tarred portion the roads were slippery. We had hardly driven five minutes when, just past the ancient Pandrethan temple, a truck moved onto the rough in order to give way to our car. Suddenly, to my horror, I saw it skidding violently towards us. Instinctively I pressed my left foot forward to protect my right hip. The truck hit us with a sickening thud. I looked down; my left leg had broken between the knee and ankle and was bent ninety degrees. There was also a sharp pain in my bad hip. I thought that I was having a nightmare and would awake to find myself back in bed. It was in fact a nightmare, but a waking one.

How I was extricated from the car, carried into the military hospital a few hundred yards away, X-rayed, put immediately under anaesthesia and awoke in agony in a hospital room with my leg and hip both in plaster is something which, even at this distance of time, I find it impossible to describe. The next two days were unadulterated agony. Sheikh Abdullah came to my room in great concern, and ran his hand across my face and head in a gesture of genuine commiseration. My father, when he heard of the accident, insisted that I must be flown down to Bombay immediately. I wanted to wait for a few days until the pain subsided, but he was adamant. An Air Force DC-3 was chartered, and once again I was lifted into it and flown out. The plane stopped to refuel in Delhi where my mother, who had rushed down from Kasauli, met me in a state of shock, unable to speak. I then flew on to Bombay where my father met me at the airport. The eminent orthopaedic experts Dr Kini and Dr Mulgaonkar were called in as soon as I reached 19 Nepean

Road. By then my leg had swollen dangerously, and on removing the plaster under yet more anaesthesia they discovered it to be a mass of blisters. Had I delayed coming down my leg might well have had to be amputated.

So once again I was back, not merely to square one but to minus one. In America only my hip was involved. Now both bones of my left leg were broken, and my hip had again cracked under the impact of the accident. Both lower limbs were in plaster and I was in agony, convinced that destiny was determined not to let me live a complete life. Mercifully, it is not easy to recall pain, because in this second disaster I perhaps underwent more than comes to an average person during an entire lifetime. For weeks I was in plaster, but the bones would not heal. My mother came down from Kasauli, and the big question was what should be done about the wedding. Asha's uncle, General Bijaya Shumsher, who was later Nepalese Ambassador in New Delhi, came to Bombay to see how I was. Obviously the January date would have to be postponed, but for how long depended upon my condition. Dr Kini and Dr Mulgaonkar tried several manipulations under anaesthesia but none of them worked, while the blisters on my broken leg took several painful weeks to heal under prolonged antibiosis. After the exhilaration of the short summer of freedom in Kashmir, I was firmly back in the dumps.

Meanwhile, of course, events in the country were moving onwards. On 26 January India's new Constitution was adopted. It was a day of great rejoicing; India had finally emerged as a full-fledged sovereign republic, and Dr Rajendra Prasad took over as the first President of India from the outgoing Governor-General, C. Rajagopalachari. In Bombay there was an unprecedented procession which lasted many hours. Even lying on my sick bed I felt a surge of happiness in becoming for the first time a citizen of a sovereign, democratic India. Happily I belonged to a generation that had not actually wielded power in the old princely order, so that while I did have a special attachment to my State this was far outweighed by the pride of being an Indian. My only regret was that on both crucial occasions, Independence Day in 1947 and now Republic Day in 1950, I should have been prostrate on my back, unable to participate actively in those historic events.

When my bones stubbornly refused to join, the doctors finally decided that they had to operate. The operation was performed and I came round with a metal strip and six screws in my left leg to balance the metal pin in my right hip. My mother had locked herself into her puja room on the day of the operation with instructions only to be informed after it was all over and I had regained consciousness. It transpired that it was just as well that the metal strip had been used, as the broken bones were proving unusually recalcitrant and I could have spent another year in plaster without them joining. Nature's ways are marvellous, but some help from intelligent surgeons can be invaluable in certain situations.

While I was recovering from the operation, I was also in touch with the State leaders, Sheikh Abdullah and his Deputy Prime Minister, Bakshi Ghulam Mohammad, a very different sort of person who was to play a crucial role in the political developments that lay ahead. Sheikh Abdullah wrote to me from time to time, including from New York where he had gone a second time as member of the Indian Delegation to the United Nations when the Kashmir dispute was still a live issue. Meanwhile, a new date for my wedding had been fixed in consultation with the doctors, who thought I should be in a position to walk again six weeks after the operation. My father was determined that the auspicious date should not upset his racing schedule, and the Pandit, suitably warned, came up with a convenient *mahurat*, 5 March. Asha's grandfather was still Prime Minister of Nepal, the last of the Rana rulers whose dynasty had virtually ruled Nepal for a century. It is a curious coincidence that our two dynasties were founded at almost exactly the same time by Maharaja Gulab Singh and Maharaja Jung Bahadur respectively, and that they came to a close almost simultaneously a century later.

Once the date was set, Asha's parents—General and Rani Sarada Shumsher—along with their five children came down to Bombay and settled in at Kutch Castle, a palatial structure then standing on Nepean Sea Road. Normally we would have gone to Nepal with a bridal procession, but as I was in no condition to travel it was decided to have the wedding in Bombay. While my parents went across to meet Asha and her family, it was considered improper that we should meet again before the actual marriage. A formal betrothal ceremony was held in the staircase hall of Kashmir House, when a representative of the Ranas along with two Nepali Pandits came over with the auspicious saffron *tilak* and the engagement ring.

Gradually I began to get out of bed and started moving around with the help of the walking sticks I had brought from America. As the wedding date approached, the house began filling up with guests and relatives. If the wedding had been in Jammu the whole city would of course have been mobilized, but in Bombay my father restricted the invitees to close relatives only. With his characteristic thoroughness he worked out all the details of the various functions preceding and following the wedding, while my mother concentrated mainly on the shopping.

Not surprisingly, it was with a mixture of pleasure and apprehension that I awaited the big event. Having been brought up an only child and gone to a boys' school, my contact with girls had till then been marginal, and for three years before the wedding my broken bones had kept me effectively demobilized as far as the social circuit went. By the end of February I could just about walk without support, but I was most uncertain of getting through the prolonged wedding ceremonies without collapsing midway. As it turned out, a feudal accessory came to my rescue. The sword, which it is obligatory for the bridegroom to carry at a Rajput wedding, doubled as an effective walking stick without causing undue embarrassment.

A day before the wedding the *Swayamvara* ceremony was held, a celebrated and ancient Hindu tradition where the bride 'chooses' her husband from among an array of suitors. Through the ages the position of women in Hindu society fell steadily with disastrous consequences, but the symbolic ceremony was retained by several communities, especially the Ranas of Nepal. Happily for me, the choice was also whittled down to one, rather like the 'free vote' in many countries when there is only a single candidate in the field. For the *Swayamvara* I wore a long brocade coat, a turban with an emerald head ornament, and a jewelled sword. My father and I drove to Kutch Castle in a huge open car—a Packard belonging to the Thakore Sahib of Dhrol, one of the fattest men in the world. It was the only car I could get into with comfort, and so came in most handy. We were received by General Sarada and General Singha, and escorted into a large hall where Asha—so heavily draped in red brocade that for a moment I thought she was a large bundle of clothes—sat on a chair across the floor from me. Then, to the chanting of Sanskrit verses, Asha got up and garlanded me—the symbolic act of choosing—after which I did the same. Asha was then carried off in a decorated palanquin by Nepali servants and maids, her face remaining fully hidden during the whole ceremony.

The next morning, 5 March 1950, I awoke to the strains of Bismillah Khan and party playing the auspicious *shehnai* in the garden. At the best of times getting married was a bit of a gamble, and here I was going to marry a girl of thirteen whom I had only met once for half an hour in the presence of two sets of parents. It was with some trepidation, therefore, that I donned my wedding clothes—a light pink silk shirt, tight pyjamas, another long brocade coat, a red turban and a set of diamond jewellery. My father was also dressed up for the occasion, wearing his fabulous emerald and diamond crown with a matching sword, along with his medals of which he was always very proud; and all the other members of our party including several former ruling princes also wore brocade. The procession was led by an army band, followed by 24 uniformed servants each carrying a silver tray containing jewellery, clothes, dry fruit and sweets as our presents to the bride, then my father with the other princes, relatives and members of the staff on foot. Finally came my open car, tastefully decorated, in which I sat behind in solitary grandeur. The drive from 19 Nepean Road to Kutch Castle on Nepean Sea Road took just over half an hour. At the entrance to Kutch Castle we were greeted by another band and finally came to a halt in front of the main building, where my father and Asha's father formally embraced—the *milni* ceremony—and other members of the two parties greeted each other.

The various elements of the wedding ceremonies reflected the chivalric tradition of the Rajputs. In medieval days travel was difficult and hazardous, and the danger from hostile clans and bandits was ever present, hence the swords for self-protection. In those days the formal meeting between the heads of the two clans being united in wedlock—exogamy being a strict requirement—was a significant event. All Rajputs trace their ancestry to one of the major clans of Rajputana. Thus our family belongs to the Kachwaha clan whose head was the ruler of Jaipur, while the Ranas traced their ancestry to the Sisodhias of Udaipur. The whole concept of the heavily armed bridal procession coming to the bride's home and returning to their fortress with her safely in tow reflects the compulsions of a bygone age, and yet provides a colourful spectacle that continues down to the present day.

After the *milni* my father and the rest of the party were seated under a spacious and colourful canopy specially erected for the occasion, while I was taken off to the *puja mandap*, the enclosure where the religious ceremonies were to take place. After some preliminary prayers to various deities representing different aspects of the all-pervasive divinity, Asha was carried into the enclosure on a palanquin by eight servants preceded by eight Nepalese maidservants dressed in red sarees and carrying the auspicious whisks made of white horse tails. With her arrival the marriage ceremonies got started, lasting over an hour, with Asha, her parents and myself being the main participants. The actual giving away of the bride is done by the father and mother jointly, unlike in the Western tradition where it is done only by the father.

After this comes the heart of the ceremony when the bride and groom walk seven times round the sacred fire, which symbolizes the divine witness to the sacrament of marriage. Asha walked so fast that all I could do was keep up with her, using my trusty sword as a walking stick, and my father had to get up and whisper to her to walk more slowly. After the circumambulation we went through a charming little Nepalese ceremony where the newly-married couple have to sit on a bed and play a game of dice with long, ivory pieces. Even before he announced the result I could see from the triumphant look on the face of the Nepali head priest that Asha had won hands down.

After the wedding ceremony was over all of us drove back to Kashmir House, leaving Asha at Kutch Castle. The next day was the auspicious time for the bride to enter her new home. I came back to the Castle and, after more ceremonies, we finally drove off together to 19 Nepean Road where we were greeted by my parents and a vast assembly of guests. Before we alighted a huge black goat was lifted over our heads to avert the evil eye, and then we walked up steps into the hall and into a glittering array of tiaras and jewellery worn by my mother and other ladies attired for the occasion. The Governor of Bombay happened to be no other than Raja Maharaj Singh, who many years earlier was for a brief while my father's Prime Minister. He and Lady Maharaj Singh were also there, as well as the Chief Minister of Bombay, B. G. Kher. The Home Minister of the State, Morarji Desai was not present, which was perhaps just as well considering the fact that champagne flowed like water and that, by the time the party broke up in the early hours of the morning, ninety per cent of the guests were in high spirits.

When we finally got up to our room it was almost two in the morning. After we had taken off all our jewels and brocades we realized that we were, in fact, very young strangers. And thus it was that we became man and wife.

Chapter 10

The next few weeks were a hectic round of parties given by us, the Nepalese and mutual friends. At some of these my father would get me to sing. I had kept up an interest in music, and despite the break of several years was able quite easily to recall the ragas and songs I had learnt as a boy. My nineteenth birthday was four days after the wedding, for which Asha gave me a Japanese transistor radio which was quite a novelty at the time. She had turned thirteen only a couple of months before we were married. Looking back, it seems incredible how young we were and how, despite difficulties in adjusting to the new situation, we were able to grow into a happy marriage. For Asha, of course, the change was greater. From one of five children living in a huge joint family she suddenly found herself married to a virtual stranger, in a family with a single son and speaking a language which she hardly knew. On top of this, the attitude of my mother, who was always very possessive towards me, was not nearly as understanding towards her as it should have been. For me, too, the change was a major one. Having been an only child, brought up virtually alone, the entry of a new person into my life marked a fresh dimension for which I was emotionally and psychologically not fully prepared. To be successful any marriage needs a great deal of mutual adjustment, but in our case this was even more necessary than usual.

For several months my convalescence and marriage had taken my mind away from the political problems in the State, but the time now came when I had to get back to the job of being Regent. It was decided that we would leave Bombay on 28 April. Bombay was warming up, and my mother, who never liked the heat, decided that she would go back to Kasauli. My father was upset at this, as he had perhaps assumed that after the wedding she would stay on. She decided to leave three days before we were to go. I still remember her coming down the stairs of Kashmir House the evening her train was leaving, and bowing low to my father with folded hands as she said goodbye. What I did not know at the time was that they were never destined to meet again.

Asha and I, along with our staff and servants, left Bombay in a chartered DC-3 early in the morning. We were in Delhi at ten, where we broke journey to call on Sardar Patel who was then at 1 Aurangzeb Road. He was very ill, but he made a special effort to spend some time with us and seemed genuinely pleased to see us. Considering that in a way he was instrumental in arranging our marriage, we thought the least we could do was to stop over and pay him our respects. Jawaharlal was not in Delhi at that time, so we could not see him. We had lunch at the Nepalese Embassy with the Ambassador, Asha's grand-uncle, and then flew on to Jammu which we reached exactly at four. There a truly massive reception awaited us. Apart from Bakshi Ghulam Mohammad and other officials, the whole of Jammu city seemed to have turned out for the welcome and the huge crowds were extremely enthusiastic and affectionate.

The day after we arrived Sheikh Abdullah came down from Srinagar and hosted a formal reception in our honour in the Mandi Mubarak, the old city palace, at which he made the graceful gesture of presenting me with a diamond *sarpech* for my turban on behalf of the State. There was also a ladies reception and endless callers at the palace. Asha took to the whole *tamasha* like a fish to water, and although she did not know a word of Dogri she seemed to manage rather well with whatever Hindi she had learnt. After a fortnight in Jammu we flew up to Srinagar and I was back at Karan Mahal, this time with a lovely child-wife in tow. Asha was thrilled with the Srinagar house and the glorious weather after the heat and bustle of Bombay and Jammu.

Gradually we began settling down to our new life together. It took me some time to realize that I was no longer simply an individual and that there was someone else who would constantly be part of my life and activities. Looking back I find it extraordinary how quickly Asha adapted to the new situation. She began to pick up Hindi, and to coach her in English I engaged Mrs Hanley, an Indian lady brought up by missionaries in Kashmir and at that time married to the author, Gerald Hanley. Asha showed some interest in painting, so we got Mrs Lilian Purbi, an Englishwoman married to a retired

Kashmiri official, to come over and give her lessons. For my part, I resumed contact with Professor Chaku. One day, while we were discussing some points in political science, he suggested that it might be a good idea if instead of diffuse and desultory studies I were to take up the B.A. course of Jammu and Kashmir University and appear as a private candidate the next year. The idea immediately enthused me, and I started studying with political science, economics and English as my subjects. I owe a debt of gratitude to Professor Chaku for this suggestion, because it enabled me to resume my academic interests and laid the foundations for future specialization. It was probably unique that I was studying to graduate from a University of which I was myself Chancellor.

In addition to the B.A. course, Pandit Parmanand began to come again to teach me Sanskrit. Thus, along with official duties which included making speeches at opening ceremonies and entertaining an unending stream of callers, we launched on a fairly stiff academic programme. I resumed reading my favourite thinkers—Bertrand Russell and Aldous Huxley— and was deeply impressed at the masterly way they took up meaningful ideas and dissected them with clarity and imagination. It struck me then how superior the realm of ideas was to the realm of men; none of the grasping corruption, the crass opportunism, only dazzling concepts moving through the clear mind like a flight of swans on the trackless path of the unknown. I also discovered Plato at this time, and was enraptured. The *Symposium* remains among my favourite books, and though I later turned towards Vedanta I continue to value my early encounter with the great seer of the Western world. In his *Republic* I came across the concept of the philosopher king, and once again my youthful imagination was aroused at the possibility of emulating that great ideal.

Meanwhile political developments continued. On 13 July 1931, during the period when Sheikh Abdullah's Muslim Conference had first launched its anti-Dogra movement, an unfortunate incident of an attack on Hindu shopkeepers had taken place resulting in communal riots and firing by the Dogra police in which several Kashmiris were killed. Since then the National Conference observed this as Martyrs' Day. In 1950 Sheikh Abdullah wanted to make some dramatic announcements on that day in order to gain maximum political mileage, and a month before he sent me for signature two proposals regarding abolition of Jagirdari (the feudal land-grant system) and of landlordism, without any compensation at all. I was in broad sympathy with the measures proposed, but obviously they would adversely affect large sections of people, particularly the non-Muslim population of the State. Vishnu Sahay had come up to Srinagar on a visit, and I thought it would be appropriate to refer the matter to him so that the Government of India could apply its mind to the implication of the proposals. When Sheikh Abdullah reminded me about the papers I informed him about what I had done.

This elicited an angry memorandum in which he roundly took me to task for having referred the matter to the Secretary for Kashmir Affairs. He wrote:

> It appears that the obvious limitations of the Constitutional head of the State have escaped the attention of Shri Yuvaraj. The action taken by Shri Yuvaraj is entirely opposed to the spirit of the Proclamation dated 5th March 1948 under which the present Government was set up. It has all along been understood that the Ruler or his Regent will act strictly as the Constitutional head of the State and it is on this distinct understanding that the present government hold their office.

He went on, significantly, to add:

> It is also clear that the subjects in regard to which the submissions were made did not relate to a matter in which the State had acceded to the Dominion of India and, therefore, in the existing constitutional set up the Government of India has no jurisdiction in this matter.

This was a fairly stiff broadside, but I decided not to buckle under. Vishnu Sahay had earlier flown down to Delhi with the two proposals, so I sent an ADC down post-haste with Sheikh Abdullah's angry note and asked Sahay to consult the 'high command', which in effect meant Jawaharlal, Sardar Patel and Gopalaswami Ayyangar, and let me have a draft reply. He came back the next day with a draft which I appended to the Sheikh's memo and sent back to him on 12 July. It read as follows:

> I appreciate the Constitutional position under the Proclamation of March 5, 1948. None the less it seems to me that these very far reach-

ing proposals, which will drastically affect the economy of very large sections of the population and for the sanctioning of which there is no properly functioning legislature at present, should, in the existing delicate political situation, be first examined in concert with the Government of India, and that we should not rush this legislation without satisfying that Government.

I would ask the Ministry to consider this aspect of the matter.

Despite this, on the next day Sheikh Abdullah made a long speech at Lal Chowk in Srinagar in which he went ahead and announced the measures, despite the fact that he had been called to Delhi to discuss the matter and that the measures were not legal without my signature. This was the first of a series of differences of approach and opinion between him and the Government of India that grew over the next few years and culminated in the dramatic events of August 1953. From the very beginning I looked upon my role as Head of State from one single angle—how best I could help to safeguard the national interest in the State as reflected in the positions taken from time to time by the Government of India under Jawaharlal Nehru. Evidently my refusal to sign these proclamations came as a rude shock to the Sheikh, as he had expected me to be merely a rubber stamp. Thus began a divergence of approach and slow evaporation of goodwill between us. Although the conflict has often been portrayed as a hangover of the old Kashmiri–Dogra animosity, the fact is that, although proud of my Dogra heritage, I tried to function not so much as a Dogra but as an Indian. Indeed, it would be correct to say that the basic difference between Sheikh Abdullah and me lay in the fact that while he looked upon himself as a Kashmiri who happened to find himself in India, I considered myself an Indian who happened to find himself in Kashmir.

There was also the still undetermined question of the exact status of Jammu and Kashmir. I have recorded how my father was vaguely attracted to the idea of an independent State, but was overwhelmed by rapidly changing circumstances. Despite all that had happened subsequently, including the accession, the war, the ceasefire and the United Nations intervention, the Sheikh too was by no means averse to the concept of independence. This became amply clear over the next two or three years, and the now public State Department papers which speak of the U.S. Ambassador Loy Henderson's discussions with him in the summer of 1950 only confirm what is well known. Before he could move in that direction, however, Sheikh Abdullah realized that he had to eliminate all vestiges of Dogra rule and gather absolute power in his own hands. It was to this end that he turned his attention with single-minded ruthlessness, aided and abetted by a host of aides and advisers led by Mirza Afzal Beg, the Revenue Minister who played Todar Mal to his Akbar.

In this murky situation my only strategy was to keep in close touch with the Government of India and to maintain personal contacts with Jawaharlal. Soon after the proclamations episode, Asha and I went down to Delhi. Asha drove up to Kasauli to visit my mother, while I stayed for a few days at Teen Murti House as a guest of Jawaharlal. Padmaja Naidu was also living there at the time, and she was always great fun, immediately putting me at ease by striking up an amusing conversation. Indira Gandhi was hostess, courteous but rather withdrawn and uncommunicative. Rajiv and Sanjay were very small at the time, and I recall walking with Jawaharlal in the back lawn of the house while he showed his grandchildren the giant pandas that had recently come as a gift from China.

Sardar Patel had only a few months left and it was not possible to meet him frequently. I did, however, start regular meetings with the President, Dr Rajendra Prasad. A kindly and cultured man, he always received me with affection and listened with interest to my reports on the situation in the State. It was clear that his approach was closer to the Sardar's than to Jawaharlal's, and he was particularly concerned about the difficult position in which the Hindus of the State were placed. He seemed to share the Sardar's deep reservations about Sheikh Abdullah, but was far too gentle to express them unequivocally. It was Dr Rajendra Prasad who came up to Srinagar later that season to address the second convocation of the University at which I once again presided.

Although the proclamation problem regarding Jagirdari and land reforms had been sorted out after Sheikh Abdullah grudgingly made some amendments on the advice of New Delhi, in which Sardar Patel even in his last weeks took an interest, it soon became clear that this had been only the opening gambit in his carefully planned strategy to consolidate his position and gain overriding constitutional legitimacy. It must be remembered that at the time the Kashmir issue was still a live one at the United Nations, where Jawaharlal's overriding idealism and sense of fair play had made the whole matter embarrassingly complicated for India. Using the leverage of the proposed plebiscite, Sheikh Abdullah continued his relentless drive against my father. Not content with having had him exiled from the State, he now began to press for the formal abolition of the ruling dynasty. And the mechanism he chose for this was the concept of a Constituent Assembly for the State. The snag, however, was that such an Assembly could not legally come into existence without my approval.

The concept of the Constituent Assembly to frame a constitution for the State was part of a broader political strategy to make the question of plebiscite redundant. Although on several occasions Jawaharlal reassured the U.N. and Pakistan that India stood by her earlier commitment, it was clear that if a Constituent Assembly did meet and reaffirm the State's accession to India it would have an impact on public opinion abroad. For Sheikh Abdullah, of course, it also provided an excellent opportunity to strike a final blow at the Dogra dynasty and thus fulfil his life's most cherished ambition. During the whole of 1950 there were discussions about this between the State Government and the Union Ministry for Kashmir Affairs. Although my father had designated me as Regent, he had not abdicated and still remained legally the ruler. It therefore became incumbent upon the Government of India to consult him about the Constituent Assembly, and on 30 November 1950 Vishnu Sahay, Secretary for Kashmir Affairs, wrote to him enclosing a draft proclamation to be issued by me and asking for his comments. On 10 December my father wrote a long letter to Sardar Patel in which he made several points which he summarized as under:

1. That the Proclamation, with the object and spirit of which I wholeheartedly agree, should be issued by me as Ruler who is the properly constituted authority in law to promulgate it and not by my Regent.
2. The powers and functions of the body intended to be constituted should be express, well defined and accurately worded and should

exclude from the purview of their inquiry and consideration matters not expressly entrusted to them.

3. They should report to the authority that constitutes them, that is the Ruler who shall seek the advice of the Parliament of India in the matter.

Seen in the light of the remarks made above it will be evident that the Proclamation as drafted if issued would lack proper sanction and will be ultra vires as its various clauses conflict with each other and with the accepted constitutional principles and in actual working of the scheme envisaged will present serious difficulties defeating the very object which it is intended to achieve.

He also wrote to me saying that as he was taking the matter up with the Government of India I should not sign the proclamation until the matter had been clarified.

Soon thereafter Sardar Patel passed away, leaving a vacuum in national life that could never really be filled. On 25 January Vishnu Sahay wrote to me enclosing a copy of the proclamation and suggesting that when I received a 'submission' from Sheikh Abdullah in this regard I should sign it. This submission came on 27 January and read as follows:

The present State Government has always been committed to the position that it is for the people of the State to draw up a constitution of the State, in whatever way they like, and that for this purpose a Constituent Assembly was to be convened at the appropriate time. The Ruler also supported this idea and in the Proclamation issued on 5th March, 1948 provision was made for convening a National Assembly when the normal conditions were restored. In October, 1949 the Government, after taking note of the present conditions and in consultation with the Government of India, decided that the time had come for the setting up of this Constituent Assembly and in pursuance of this decision steps were taken for preparation of the electoral rolls in the State. It was, however, considered desirable that a Proclamation should be issued for convening a Constituent Assembly in the State and that clauses (4) to (6) of the Proclamation, dated 5th March, 1948, superseded, as they did not meet the present requirements. The terms of this Proclamation and the powers and functions of the Constituent Assembly have been the subject of discussion and correspondence with the Government of India for a long time and the Prime Minister feels gratified that in the end a complete agreement has been reached between the views of this Government and those of the Government of India in this respect. The enclosed draft Proclamation has been prepared jointly by the two Governments and has the full concurrence of the Government of India. The Government of India desire that this Proclamation should be issued at as early a date as possible. It is, therefore, requested that the draft Proclamation may be returned forthwith after Shree Yuvaraj has affixed his signature to it.

S. M. Abdullah
PRIME MINISTER
27.1.1951

These developments put me in an acute dilemma. On the one hand was the express direction from my father that I should *not* sign the proclamation, and on the other both the Government of India and Sheikh Abdullah were urging me to sign it. I decided to fly at once to Delhi where I met Gopalaswami Ayyangar, who had taken over as Minister for States, and explained the position to him. On my return I made the following note upon Sheikh Abdullah's submission:

PRIME MINISTER

As you are aware, I derive my powers from His Highness' proclamation dated 20th June 1949, appointing me Regent.

The draft proclamation which you have now put before me is a very important document, involving, as it does, far reaching consequences to the future of the State. The Ministry of States had sent a draft proclamation to His Highness for his comments. His Highness sent his comments thereon to the then Hon'ble Minister for States (the late Sardar Patel), and at the same time directed me not to append my signature on any proclamation on the subject without his express approval.

You will, therefore, appreciate that as I have received no such approval from him yet, I am at this stage precluded from signing the document you have submitted to me. I therefore suggest that you ask the Government of India to pursue this matter with His Highness.

Karan Singh
REGENT
30.1.1951

This put the ball back in the Government of India's court. Meanwhile my father took legal opinion from a leading firm of lawyers, Kanga and Company, which formed the basis of a detailed memorandum that he later submitted to the President. The opinion upheld the view that he remained Ruler who alone had the power to sign a proclamation for the setting up of the Constituent Assembly. Unfortunately the question involved was not one of legal niceties but the harsh realities of political power. V. P. Menon met my father in Bombay, and then on 5 April Gopalaswami Ayyangar wrote a letter to him in which the position was bluntly stated as under:

> ... Developments have, however, since taken place both in the State and at Lake Success which make it imperative that the issue of this Proclamation is not delayed any longer. The Government of India are committed to the convening of a Constituent Assembly, the preparations for which are in active progress in the State. That Assembly will be held whether the formal Proclamation issues or not. In the view of the Government of India it must be convened, if both their commitments to the people of Kashmir and their stand at Lake Success are to be implemented, in spirit and in the letter. From the beginning, they have held that this Constituent Assembly should be called under the provisions of the Constitution of India and that this should be done, from both a tactical and constitutional point of view, on the authority of a Proclamation issued by the Head of the State. The draft of the Proclamation has been agreed between the Government of India and the Government of Jammu & Kashmir. No purpose will, therefore, be served by any act of Your Highness which holds up the signing and issue of this Proclamation by Shri Yuvaraj.
>
> On neither of the two matters about which I can understand your entertaining apprehensions, namely, the continuance of the accession of the Jammu & Kashmir State or of part thereof to India and the connection of the Headship of the State with your dynasty, no *final* decision should be taken by the Constituent Assembly to be convened. They are essentially matters which could be decided only as a matter of agreement between the Government of India and Parliament on the one side and the Government of Jammu & Kashmir and the State Constituent Assembly or legislature on the other. The Government of India will, no doubt, at the proper time take the decision on these matters, which, I need hardly assure you, would be essentially just from the standpoint both of your dynasty and the people of the State. You have obviously to put your trust in the people of the State and the Government of India in respect of this matter. I hope, therefore, you will immediately lift the ban which you have placed on Shri Yuvaraj affixing his signature to the agreed Proclamation and which naturally has placed him in great embarrassment. ...

This persuaded my father to fall in line, and he wrote to me formally saying that he lifted the ban and left the matter to my discretion. A few days thereafter I received a letter from V. P. Menon urging me to sign the proclamation, which I did on 21 April. Meanwhile, Sheikh Abdullah had evidently not taken kindly to my earlier refusal to sign. In order to keep up the pressure he launched a bitter personal attack on my father and me at a public meeting a few miles from Jammu on 9 April. He again accused my parents of instigating communal riots in 1947 and said that if I was not careful I would also have to go the way he went. This utterly unprovoked attack came as a shock to me, and I immediately wrote to him and sent a copy to Jawaharlal who had just returned from Kashmir, with a covering letter in which I added: 'I am very deeply hurt at Sheikh Sahib's making such remarks in public, particularly as they were not based on facts and were consequently highly misleading.'

The Sheikh responded to my letter with a long five-page epistle in which he made it clear that he expected me to act virtually as a captive of his Government, even to the extent of consulting him before making public appearances. This letter revealed that although he might tolerate me as Head of State, he was incapable of overcoming his pathological aversion to my family. At the time he was riding the crest of the wave, however, and there was little I could do except to lie low for the time being. I kept up my contacts with Jawaharlal. Every time I went to Delhi I would meet him, and he would talk at length on many matters. Kashmir was always uppermost in his mind, and many of his letters to me at that period deal in detail with various internal and international developments relating to the State. It seems that he had developed some affection for me, because he always seemed genuinely pleased to see me. In a letter of 13 May 1951 to my mother he ended by saying: 'I have seen Tiger from time to time and have grown to like him very much. He is a fine young man and I think his good qualities will enable him to face difficulties. I shall certainly help and guide him in every way.'

I quote this because it has a bearing on developments that subsequently took place, and also because it shows that, despite Sheikh Abdullah's special position, it was not as if the other point of view was not being listened to by Jawaharlal. D. P. Dhar also had direct access to Jawaharlal, and between us we were able to counteract to some extent the belligerent Muslim Kashmiri chauvinism represented by the Sheikh and the hard-liners in the National Conference led by Mirza Afzal Beg. Jawaharlal's psychology in this regard was interesting. If he rejected a person's bona fides, then any view or advice from those quarters would at once be suspect. Thus he had a particular dislike for the Praja Parishad leaders in Jammu, because he thought they were communalists and that their actions only jeopardized India's position *vis-à-vis* Kashmir. But if he trusted someone—and he was loyal to a fault as far as his friends were concerned—he would give a careful hearing to their views even if they were not in tune with his own thinking.

Jawaharlal visited Srinagar early in June that year. Evidently Sheikh Abdullah had spoken to him about our brush a few weeks earlier as a result of his Jammu speech, and had suggested that a Secretary should be appointed on my staff. Jawaharlal wrote to me about this from Chashmashahi House where he always liked to stay. The house belonged to us and it was here that my mother and father had been married in 1928. Subsequently it was used as a sort of special guest house, and Sir Tej Bahadur Sapru would stay there for several summer weeks along with members of his family. Later it was the residence of Swami Sant Dev, and after 1947 it became the favourite residence of Jawaharlal. The house commands a magnificent view of the Dal Lake, the Shankaracharya and Hari Parbat hills and the higher mountain panorama behind them. The sunsets are particularly spectacular when viewed from here, and Jawaharlal would often sit quietly on the veranda watching the sun sink below the horizon. It is now the Raj Bhavan.

I was meanwhile continuing my preparations for the B.A. examination, while Asha had also begun studying English and Hindi, as well as learning how to paint. Our life was beginning to fall into a pattern. In the mornings we would study with tutors for about two hours, after which I would have some formal meetings with officials or visitors. Two or three times a week we had some people over to lunch, including visiting dignitaries such as foreign ambassadors, United Nations personnel and the local gentry. Asha was beginning to pick up conversational English but was still a little shy, so I had to carry on conversation for both of us, which was often amusing. We would sit opposite each other at the middle of the dining table, with guests ranged on either side according to their listing in the warrant of precedence. Our table took a maximum of sixteen people, so the parties were intimate. Over the years I found these meals of real value, because a broad spectrum of people—some highly intelligent and stimulating—came from various parts of India and many other countries of the world. Always a quick learner, I picked up valuable general information and ideas from those conversations.

Apart from being interested in ideas, I found myself drawn towards education, and this led me to suggest to my father that we should donate Gulab Bhavan, our main palace in Srinagar, to the newly established Jammu and Kashmir University. I mentioned this to Jawaharlal and Gopalaswami Ayyangar, who reacted favourably. Jawaharlal wrote: 'I am very glad to learn that you intend donating the Gulab Mahal Palace to the University of Jammu and Kashmir. That is the very best use you can put the building to and I am sure it will be greatly appreciated by the public.' My father was out of India at the time, but his Private Secretary, Bhim Sain Mahey (brother of the communist leader Dhanwantari Mahey), had come to Srinagar and I had asked him to convey my views regarding this donation to my father. Evidently he did not do so, but Gopalaswami Ayyangar meanwhile wrote to my father mentioning the offer and urging him to accept it. My father was furious and wrote me an angry letter. I wrote back, laying out the factors which had impelled me to suggest the gift and summed this up by saying:

> Considering all these factors—a fast deteriorating building which is ideally suited to be a University and which is lying neglected and unused, and the fact that our University greatly requires a suitable building and that education is as noble and as worthy a cause as can possibly be found—I sincerely feel that it would be a magnificent gesture on your part if you were to donate the building. The effect of such an act on the public would, I am convinced, be electrifying.

My father, however, disagreed and in his reply angrily said that

The Jammu and Kashmir Government have gone out of their way to harass me personally and to humiliate the dynasty in the eyes of the public. . . . I hope you will realize that the gift in the present circumstances with however lofty ideals behind it would be misconstrued as one more instance of the Ruler being proved a non-entity and of no consequence.

And there, for the time being, the matter rested. Soon thereafter Gulab Bhavan was converted into a beautiful hotel, providing tourists to Kashmir with top-class accommodation for the first time. However, I did not forget my commitment to the University, and some years later donated a 120-acre orchard near Hazratbal (named Amar Singh Bagh after my grandfather) to Kashmir University, where the present campus stands on a magnificent site.

Towards the end of October I sat for the B.A. examination. As I was Chancellor, the University authorities offered to open a special centre at my residence, but I thought that would be most inappropriate and decided to take the examination along with other students at the Sri Pratap College centre. The only concession was a higher chair and desk than provided for others, as I had difficulty sitting at a low desk on account of my stiff hip.

On 5 November 1951 the Jammu and Kashmir Constituent Assembly met. Elections to this august body that I had summoned by proclamation had been held earlier, but the only effective opposition group—the Praja Parishad in Jammu—boycotted the elections. As a result of this, 72 out of the 75 members were elected unopposed on the National Conference ticket, and no Praja Parishad members were elected, including the veteran and widely respected leader Pandit Prem Nath Dogra. I did not attend the opening session, but for Sheikh Abdullah, of course, it was his finest hour and he played it to the hilt. In a somewhat pompous and flamboyant opening address in which, among others, he quoted Thomas Aquinas, he pointed out that, apart from framing a Constitution 'for the future governance of the country', the Assembly would declare its 'reasoned conclusion regarding accession' and also deal with the land reforms passed by the Government. He added with relish: 'Another issue of vital import to the nation involves the future of the Royal Dynasty'. After a long description from his point of view regarding the historical background of the Dogra dynasty and its performance, he launched a broadside against my father. He said:

After the attainment of complete power by the people, it would have been an appropriate gesture of goodwill to recognize Maharaja Hari Singh as the first Constitutional Head of the State. But I must say with regret that he has completely forfeited the confidence of every section of the people. His incapacity to adjust himself to changed conditions and his antiquated views on vital problems constitute positive disqualifications for him to hold the high office of a democratic Head of the State.

In the next two paragraphs, however, he came out strongly in favour of my being chosen the first Head of the State. This is what he said:

I am sure none of us is interested in a personal controversy with the Maharajah's family. In the conduct of public affairs, it is necessary that an impartial view of every individual's deeds should be taken. Our judgement should not be warped by ill-will or personal rancour. During our association with Yuvaraj Karan Singh these last few years, I and my colleagues in the Government have been impressed by his intelligence, his broad outlook and his keen desire to serve the country. These qualities of the Yuvaraj single him out as a fit choice for the honour of being chosen the first Head of the State.

There is no doubt that Yuvaraj Karan Singh in his capacity as a citizen of the State, will prove a fitting symbol of the transition to a democratic system in which the ruler of yesterday becomes the first servant of the people, functioning under their authority, and on their behalf.

Clearly this had been the compromise worked out during the long discussions conducted with Jawaharlal and other representatives of the Government of India. This was a shrewd move, on the one hand to avoid offending the Dogras irretrievably and on the other of retaining in a subtle way the constitutional authority of our family. After all, the legal and constitutional validity of Jammu and Kashmir's accession to the Indian Union rested upon the Instrument of Accession signed by my father, and now that very person was being unceremoniously pushed out by an Assembly which had been convened through a proclamation signed by me as his son and Regent! A piquant constitutional impasse was therefore caused by what both Sheikh Abdullah and Jawaharlal thought was an overriding political compulsion to totally eliminate my father from the political scene. In this unusual situation I fitted neatly into the picture and provided all concerned with some sort of a face-saving device.

Howsoever neat the solution, though, the fact remained that it would only work if I agreed to it. And this put me, at the age of just under twenty-one, in the toughest of the many dilemmas that I had faced in my short but exciting political career thus far. Subsequently I have on many occasions been forced to take difficult and major decisions, but this one remains in my memory as the most excruciating of them all.

Chapter 11

The controversy regarding whether or not I should accept the office of elected Head of State after the abolition of the monarchy rapidly gained momentum, specially in Jammu. Already outraged at the way my father had been exiled after having signed the Accession, in sharp contrast to the Nizam of Hyderabad who continued to reside in his State although his government had defied the new Indian republic and even resorted to arms against it, the people of Jammu reacted sharply to the anti-Dogra attitude and speeches of Sheikh Abdullah and his colleagues. In fact, after Independence the ruling families all over India were fading out and had lost effective power, but Sheikh Abdullah's move in making a major drama of 'abolishing' our dynasty was extremely galling to the Dogras. That I should accept to be the first 'puppet' Head of State under these circumstances was looked upon by many as a traitorous act which would lend an aura of legitimacy to the Sheikh's actions. My father himself was justifiably furious at the way he had been treated. Having reluctantly agreed to appoint me Regent, the idea that I would take over as Head of State after he had been unceremoniously abolished *in absentia* was highly unpalatable to him, and it was only after the dramatic events of 1953 that he became partially reconciled to the situation.

Looking back over the years, I can well understand his sentiments. However, several factors weighed with me at the time apart, perhaps, from some deep psychological urge to prove that I could make it on my own. One was my conviction that, whether or not I accepted the office, the Sheikh was determined to push ahead; and if I declined he would simply proceed to elect someone else instead. In fact, the name of Mahasha Nahar Singh, a Harijan member from Jammu, had already been floated in some National Conference quarters. Thus my declining would merely mean that I would have to follow my father into exile in Bombay, because the Sheikh would hardly have tolerated my presence in the State as a potential focus of Dogra resistance and discontent. Not only was the prospect of living permanently in Bombay far from attractive, it would also have meant snapping the last link of our family with the State. I had no doubt that if I left, there being no other member of the family, the Sheikh would in due course proceed to take over all our property and lands, as also the religious and family Trusts set up by my forefathers.

Apart from personal factors, however, there were the broader implications in the whole matter that affected the national interest. I realized that by staying on in the new capacity I might be able to play a useful role in safeguarding and serving the national interest in this sensitive and crucial State, and this obviously was what Jawaharlal Nehru wanted. Thus, once again, the conflict between my father and Jawaharlal became the central feature of my life. Tremendous pressure was brought to bear upon me through various members of my father's staff and senior Dogra leaders. It was pointed out that if I accepted the office it would be tantamount to selling the Dogras out to the Kashmiris; that I would go down in history as a traitor to the dynasty founded by Maharaja Gulab Singh; that if I defied my father I might well be disinherited and lose my patrimony worth many crores of rupees.

In addition to these direct assaults there were more insidious arguments. How did I know the whole arrangement was not some elaborate trap prepared by Abdullah? In the light of his pathological hatred for the family would he not, having secured the abolition of the dynasty with my connivance, seize upon some excuse or the other to turn against me soon and send me packing? Had he not only recently publicly fulminated against me in his Jammu speech, and what was there to prevent him making my position so difficult that I would be forced to quit or face total humiliation? Had the Sheikh not forced the Government of India to resile step by step from all the commitments made to my father, and would he not do the same for any fresh ones that may be made in regard to myself?

I was thus faced with an awesome dilemma, to accept or not to accept. In such a situation, my instinct was to try and gain time, and this I proceeded to do on one very strong ground. Sheikh Abdullah's Government was in the process of detailed negotiations with the Government of India regarding the new constitutional arrangements, covering not only the headship of the State but other vital matters such as citizenship, fundamental rights, the Supreme Court, financial integration, the flag, the powers of reprieve of the President and other allied matters. Jawaharlal wrote to me about these matters on 26 July 1952 in which, *inter alia*, he said:

> I am sure you will appreciate what I am writing. I need not tell you that now and later, you will be in mind and you can always come to me for advice or any help that I can give you. The best advice is to accept cheerfully and willingly the changes suggested and thereby to put yourself in the forefront of them, instead of appearing as if you unwillingly agreed to something that you disliked. If we have to do something, we should do it gracefully and thereby gain the goodwill and respect of others.

In reply I seized upon what I thought was a persuasive argument. I said:

> Before I proceed, I would like to express my deep gratitude to you for the kind and sympathetic interest you have been taking in my affairs during the last three years or so since I became Regent. I need hardly express how greatly I value your guidance and advice.
>
> As regards the question of whether I should or should not accept a five-year elected term as Head of State, I have—since your speech in Parliament was delivered about ten days ago—been giving the matter my most thorough consideration from all its various aspects.
>
> My highest ambition is to be able to serve effectively the people of my country, and any position which gives me that opportunity will naturally be welcomed. However, in the present circumstances, I feel that it is not possible for me to come to any decision until the new Constitution for Kashmir emerges in its final shape from the Constituent Assembly and receives the approval of the Government of India. I am sure you will appreciate that it is hardly possible for me to accept a position without knowing exactly and clearly the duties and functions which attach to it, the responsibilities which I will have to shoulder and the conditions under which I will have to work.

Soon thereafter Sheikh Abdullah called on me and reiterated that I must take an early decision as to whether or not I would accept to be Sadar-i-Riyasat. I tried to press my arguments but he brushed them aside, saying that the Constitution might take considerable time in drafting and he was not prepared to wait till then. I wrote again to Jawaharlal protesting against this attitude. Some days later he replied on 8 August in a three-page letter. Regarding my point of waiting until the whole Constitution was drafted he said:

I quite agree with you that this business of hustling through a part of the Constitution is neither normal nor ordinarily desirable. The proper course would have been to pass the full Constitution and then give effect to it. But this matter was discussed by us at very great length when the Kashmir delegation was here and ultimately we came to the conclusions which were embodied in the points of agreement. It serves little purpose to go back on them and renew the old argument. Much has been done which cannot be undone at the present stage and any attempt to do so would only produce further complications. Therefore, we have to accept the present position as it is. That is to say that a part of the Constitution dealing with the Head of the State, in the form agreed upon, should be adopted first.

And he ended with the following advice:

The alternative to your not accepting this will not only be bad for the State but will be of no advantage to you at this stage. You mentioned to me your desire for education abroad. A year or more abroad may do you good no doubt. But we learn more about life and its problems by facing them than by mere changes in geographical environment. You are likely to have a fair amount of time at your disposal for study. You are attracted to it and you can thus prepare yourself in many ways better in the State than if you went abroad. Most foreign countries today are full of the noise of war and its preparations and are not very pleasant places to go to except for brief visits.

I feel, therefore, that it would be the right thing for you to accept the offer in the manner I have suggested and thereby put yourself right with your people and do them much service in many ways which may not be so obvious and yet which are important. And then, after all, you retain your freedom of choice at any later moment, should occasion so demand.

I decided to go down to Delhi and discuss the matter personally. Jawaharlal as usual greeted me with great affection in his gracious drawing-room in Teen Murti. He was then at the height of his power, and spoke gently but with great conviction. He explained to me why he and his colleagues were anxious that I should accept. Evidently the Sheikh's constitutional negotiating team consisting of himself, Mirza Afzal Beg, D. P. Dhar and M. A. Shahmiri, his constitutional adviser, had taken rigid positions, because for the first time I noticed a slight hint of displeasure in Jawaharlal's remarks about Sheikh Abdullah. Although he did not say so in so many words, I gathered the distinct impression that he wanted me to be on the scene so that I could be of help if some difficulty arose in the future. When I explained to him my difficulty regarding Jammu opinion he agreed that I might invite some Jammu leaders to Srinagar for consultations. This itself was a major shift in view of his well known allergy to the Praja Parishad, and reflected the fact that he appreciated my difficulty.

It was on this visit that, at a tea party hosted by Jawaharlal in the lawns of Teen Murti, I came across C. Rajagopalachari who was then Chief Minister of Madras. Having met him once when he was Governor-General several years earlier, I went up and greeted him. 'You may not remember me,' I said, 'I am Karan Singh.' He turned round, looked at me and said, 'Of course I remember you. You have got such beautiful eyes.' It is not often that I find myself at a loss for words, but at that moment I was rendered speechless. I also called on Maulana Azad, who was the epitome of culture and erudition and whose conversation was always a joy.

On my return to Srinagar I began the process of consulting the Jammu leaders, including the President of the Praja Parishad, Pandit Premnath Dogra. After the meeting he issued a statement reiterating the demand for complete accession and concluding that 'till the new Constitution for the State assumes a concrete shape, it would be premature to give any definite opinion on this single issue of accepting the office of the Head of the State by Shree Yuvaraj Bahadur'. In fact, they were strongly opposed to my accepting, but after three days' talks I was able to bring them round to at least this intermediate position. Meanwhile, I learnt that my father had sent a long memorandum to the President, and there were press reports to the effect that he had demanded a referendum in the State upon the issue of retention or abolition of the dynasty.

Early in September I wrote to Jawaharlal reporting on my meeting with the Jammu leaders and pointing out the genuine grievances and apprehensions that they had expressed. I urged that senior Indian leaders—specially Maulana Azad, Dr K. N. Katju, the then Home Minister, and Gopalaswami Ayyangar, the Minister for Kashmir Affairs—should meet the Jammu leaders and give them a sympathetic hearing. I concluded the letter with a mention of my father's reported memorandum and, although I knew Jawaharlal would not be pleased, felt impelled to end with the following paragraph:

> I wonder what your view about this is. If it is possible to have a referendum—and I do not see why it should not be—I feel it would be a good thing, as it would give the people of the State a fully democratic method of expressing their decision as to whether they would like a member of the dynasty to be their Constitutional Head or would prefer to elect someone periodically. Thus no section or group will feel that its views have been ignored in coming to a decision on this important question. I must add that from indications both in Jammu and in the valley I feel that the result of such a referendum is by no means a foregone conclusion.

Jawaharlal replied the very next day. While accepting that the Jammu people have a number of grievances, some of which were justified, he went on to castigate the Praja Parishad for their mode of functioning. About the memorandum, he wrote: 'I have seen the long memorandum which your father has sent to the President. There is no reference in it to a referendum. The memorandum is rather an angry and tendentious document. Your father does not seem to realize at all that the world has changed and is changing rather rapidly.'

He then concluded with some comments on my mention of a referendum. He wrote:

> I am rather surprised at your reference to the referendum. This is not at all possible either from the local or from the international point of view. We have been talking about the plebiscite over other issues and even this cannot come off because there is no agreement. If the question of a referendum on a limited issue was raised, this would immediately lead to all kinds of international complications and the demand for a plebiscite immediately over the wider area. Even this referendum, it would be said, if held at all, should be over that wider area, including that part which is held by Pakistan. Within the present boundaries of the State under our control the referendum issue would, naturally, lead to bitterness and controversy and in effect tend to split up the State, regardless of the final issue. Indeed, I think that such a proposal is completely out of court in the present circumstances. Pakistan would, no doubt, profit by it, but no one else.

At about this time, I had the dramatic experience of accompanying Jawaharlal on a flight to a newly constructed airfield at Chushul in Eastern Ladakh. This was situated at an altitude of 14,270 feet above sea level, reportedly the highest airfield in the world. On the flight were Indira Gandhi, Sheikh Abdullah and several others, and we were piloted by the ace flier, Air Marshal Mehr Singh. We flew in a DC-3 at an altitude of over 20,000 feet and wore oxygen masks most of the way. The sight of the great Himalaya from the air was spectacular, the mountains looked like some vast, frozen ocean stretching endlessly in all directions. The sight of Nanga Parbat, one of the world's highest peaks, standing lone and majestic, robed in icy splendour was unforgettable. Just before landing at Chushul we flew over a huge turquoise-blue lake, and when we landed I had a strange feeling of sensory distortion. The mountains near the runway looked so close that Indira Gandhi wanted to walk up to them. However, we were told to our astonishment that they were at least ten miles away. In that rarefied air objects looked much nearer than they actually are. We stayed at the airfield for about ninety minutes while Jawaharlal chatted with the air force officers. There was no habitation or vegetation within sight, and the landscape had a stark beauty that might have belonged to another planet.

This flight to Chushul strengthened my earlier desire to pay a visit to Ladakh on my own, and soon after I decided to visit that far-flung region of 36,000 square miles which was annexed to the State by General Zorawar Singh and his intrepid Dogra forces way back in the time of Maharaja Gulab Singh. This represented two-thirds of the entire area of the State left in our possession. No member of our family had ever visited Ladakh, which was administered by a *Wazir Wazarat* (equivalent to Deputy Commissioner) during Dogra rule. It was narrowly saved from falling into the hands of the Pakistani raiders in 1947, and remains the largest Lamaistic Buddhist enclave in the world, Tibet having fallen under Chinese rule in what was one of Jawaharlal's less successful foreign policy exercises.

Asha and I landed in Leh to a rousing reception. The entire population of the town and surrounding villages had turned out to greet us, led by the Head Lama Kushak Bakula. Stepping out of the plane was stepping into another, but friendly world—the deep blue sky, the clear, rarefied air, the Ladakhis in their traditional robes and the women with their gorgeous turquoise headpieces. We were in Leh for three days, during which we visited the leading Buddhist monasteries and made offerings and prayers there. Some of the statues of the Buddha and Bodhisattvas were stunningly beautiful, and the Lamaseries were decorated with freshly painted wooden furniture and hung with priceless scrolls or *thangkas* illustrating various Buddhist deities. Hundreds of white metal cups held oil wicks which imparted a soft, mellow glow to the shrines and this, combined with the deep, continuous chanting by the monks, created a strangely hypnotic effect. I remember in particular the huge hollow metal image of the Buddha at the Shay monastery, three stories high and so large that it was used in earlier times as a grain reservoir for the entire town of Leh.

We had taken with us several thousand yards of cloth for distribution so that every inhabitant of Ladakh could get enough for a coat. It was touching to learn that these pieces were treated with reverence and affection by the people even in far-off villages as tokens from the Dogra ruling family. Quite evidently time had removed any traces of bitterness and left in its place a genuine regard for our family among the Ladakhis. We received several colourful presents, including a set of Ladakhi clothes in which I was photographed. Many years later, when I revisited the Lamaseries I was surprised to see this photograph in almost all of them. Asha was fascinated by the turquoise-studded hair ornaments worn by the Ladakhi women. From Leh we flew to Kargil, the second town in Ladakh and the centre of the area inhabited by Shia Muslims. The Ceasefire Line passed very close to Kargil, and the army here was more in evidence. The brigade headquarters where we stayed were situated on the banks of the Indus. It struck me as curious that although the very name of our country and its predominant religion are derived from the Indus, the only place where this great river now flows through Indian territory is Ladakh.

During my stay in Ladakh I had some serious discussions with Kushak Bakula and his colleagues. Even more so than in Jammu, the Ladakhis were feeling uneasy and insecure under the Sheikh's administration. Forming as they did a distinct cultural entity, they felt that their position in the new dispensation, with only two members in the State Assembly, on the basis of population, was extremely precarious and made them totally subordinate to the Kashmiris. They suggested that a Statutory Advisory Committee should be set up to which it should be obligatory for the State Legislature to refer all legislation vitally affecting Ladakh. They also urged that, instead of leaving them at the mercy of the Sheikh's Government, an administrator should be sent from the Centre.

On my return to Srinagar I wrote to Jawaharlal enclosing a note on my impressions regarding Ladakh and suggesting measures to improve the situation. Urging him to take early action in the matter I pointed out that 'Ladakh is strategically very important, more so for India as a whole even than for the State, and it will be a tragedy if for lack of an accommodating spirit the people of Ladakh remain unhappy and discontented, and thus an easy prey to all sorts of exploitation, both communalist and communist.' As usual, Jawaharlal replied within a week. He said that 'the real difficulty about Ladakh is its terrible economic backwardness', and concluded his letter with the remark that 'we should like to make a road to Leh from Kulu Valley, but this is a very expensive undertaking and for the present it is difficult to take it up. That road will open out the mineral deposits of Ladakh and will bring a certain measure of prosperity to that region.'

It has been my experience that an outer crisis often triggers off an inner response which may appear totally unconnected with the visible trend of events. While all these political events were in progress, a process of inner development was also going on. I began at this stage to develop a strong predilection for spiritual ideas. Edwin Arnold's *The Light of Asia*, that great poem on the life and teachings of the Buddha, had a deep impact upon me. At about this time I went out to Dachigam and shot a bear while it was eating mulberries on one of the trees. It fell to the ground shrieking pitifully like a child, and lay there for long before it died. Those cries haunted me for weeks, and as a result I decided to give up shooting and fishing. Despite my association with these sports from childhood, I have never since indulged in either of them. It was evidently the same influence that led me privately to record a signed note on

1 January 1952 to the effect that my long-range aim was 'to achieve spiritual enlightenment and peace, and to successfully spread the message throughout the world so that the world which is rushing headlong into the dark and frightful abyss of destruction along the road of hate, envy, cruelty and fanaticism can be saved and put on the broad and clear sunlit path of peace, happiness, love and universal prosperity'. I was just under twenty-one when I wrote this, and though the impulse remains with me, I have over the years begun to realize that saving the world is not quite as simple a task as it might have appeared then.

Meanwhile my examination results had been announced. I passed the B.A. (missing a first division by six marks) and thus became probably the first Chancellor in world history to graduate from his own University. That year the philosopher-statesman, Dr Radhakrishnan, came to deliver the convocation address. I greatly admired his writings, and the impressive manner in which he put across his ideas in crisp, concise and eloquent sentences, punctuated with quotations from the Sanskrit classics. In fact I decided then to model my public-speaking style on his. That year he gave a brilliant extemporaneous address to the convocation, which was held under the magnificent Chinar trees in the garden of the old city palace on the Jhelum, a setting far more impressive than any auditorium. As Chancellor it was my function to present the certificates to the graduates of the year, but obviously I could not present one to myself. So, after the presentations were over, I stepped down and received my own certificate from Dr Radhakrishnan.

Among the numerous books that I read during this period one deserves special mention, as it introduced me to two personalities who were to have a profound influence on my inner life in the years to come. *Among the Great* by Dilip Kumar Roy is a record of his meetings and interviews with five outstanding world thinkers—Bertrand Russell, Romain Rolland, Mahatma Gandhi, Rabindranath Tagore and Sri Aurobindo. The essay on Sri Aurobindo was particularly interesting, and though I had never met him in person I began to get increasingly drawn to his magnificent vision of a spiritualized human consciousness. More important, the essay made mention of an Englishman called Ronald Nixon who had become a *Vaishnava* sanyasi (a monk) under the name of Sri Krishnaprem. This finally led me to the feet of the most remarkable man I have ever had the privilege of meeting. I thus owe a double debt of gratitude to Dilipda, and to destiny for having placed his book in my hands at a critical juncture in my life. Sri Aurobindo and Sri Krishnaprem were to become the twin guiding stars for my inner quest, converging in the golden notes of the eternal flutist.

The whole question of constitutional developments continued to proceed inexorably. My father had been conducting prolonged correspondence with the States Ministry, but he and the Government of India were on such different wavelengths that the exercise was infructuous. My father kept insisting on his rights and the solemn commitments given to him from time to time, while the Minister for States, Dr K. N. Katju, kept reiterating the changed political conditions in the State and the broader compulsions of international politics. The Secretary in the Ministry of States, C. S. Venkatachar, called on my father in Poona in September, after which my father wrote an anguished letter to Dr Katju. Referring to the memorandum he had sent to the President of India he recalled the 'misfortunes brought about by a persistent disregard of my rights and the assurances given to me to preserve them by the Government of India, thereby weakening my position and securing very unfair and undue advantage to my persecutors', and added, 'Am I not entitled, may I ask you, to be told either that I am wrong or that the Government of India are committed to carry out a certain policy emanating from Sheikh Abdullah even though it may mean sacrificing myself, my dynasty and certain cherished principles of justice and equity? Have I lost the elementary rights of a person who considers himself aggrieved and seeks justice?'

As in some ancient Greek tragedy, however, my father found the situation implacably moving against him. There was tremendous pressure on him to abdicate, but this he resolutely refused to do. Ultimately, some other broad matters relating to his privy purse and privileges were discussed and tentative formulations arrived at. My own efforts to stall a decision in the matter were also reaching breaking point, and my hope that some agreed solution between the Government of India and my father would be worked out, thus clearing the way for my course of action, had also proved illusory. The moment of truth was fast approaching, and I realized that I would soon have to take a decision on my own responsibility and be prepared to face whatever consequences may ensue. There were no soft options, and it was becoming clear to me that the choice was really between bowing out of public life altogether and joining the ranks of former princes who patronized the Bombay race club, or of staying in the fray and fighting it out even if it meant ultimate defeat. It was also a choice between the advice and influence of two men, both powerful and strong-willed personalities.

On 30 October 1952 Jawaharlal wrote me a letter saying that the prolonged discussions with Sheikh Abdullah regarding the Head of State had finally concluded with the agreement that the President should act under Article 370 of the Constitution and vary the explanation given there regarding this office, that it was not desirable to keep the matter pending any longer, and that necessary steps would be taken around the middle of November. 'I hope that you will agree to the steps we have suggested', he wrote. 'That is the only course that is open to us now and we should not hesitate to take it. Any other course or an attempt at postponement will only lead to difficulty and trouble.' Significantly, he added, 'I am afraid your father the Maharaja has not been very co-operative. We have tried to explain matters to him and to help him as far as possible. But he appears to be totally unaware of the changes that have taken place and are taking place in the world, and puts forward some pleas which have no application in the present.'

This crossed a letter I wrote to Dr Katju in which I asked for some clarifications regarding the position and functions of the proposed office of 'Sadar-i-Riyasat', the name given to the office of Head of State under the new Constitution. I sent a copy to Jawaharlal. It is an index of his keen interest in Kashmir affairs that he responded to my letter at once, even before Dr Katju had been able to reply. After dealing briefly with the points I had raised he concluded, 'If you have to start a new relationship, it should be under as favourable auspices as possible and with goodwill. That is a stronger guarantee than anything else.'

Two days later Dr Katju also replied, dealing in detail with various points including the privy purse (reduced from Rs 28 to 10 lakhs and distributed between my father, my mother and myself), the emoluments of the Sadar-i-Riyasat, the flag, the tenure of office, and so on.

Before finally taking the plunge I decided to pay another visit to Delhi. There I called on Jawaharlal as usual. He was deeply aware of my difficulty, and spoke to me at length about the national implications of my decision and of how a person has to take a clear stand at critical junctures regardless of pressures and hesitations. He repeated some of the perplexity that he was beginning to feel in dealing with Sheikh Abdullah, and said that this made it all the more important that I should be on the scene to be able to help if some problems arose in the future. It was this final aspect that clinched the issue as far as I was concerned. Although he asked for no assurance, and I gave him none, I left Teen Murti with my mind made up to accept.

I returned to Srinagar around 12 November. On the 15th the Constituent Assembly met and elected me Sadar-i-Riyasat for a five-year term. I had turned twenty-one only a few months earlier, so the usual constitutional age limitation for Head of State was reduced from 35 to 21 years. Jawaharlal wrote me a semi-formal letter which read as follows:

My dear Yuvraj,

I write to congratulate you on the high honour that has been conferred upon you by the people of Jammu and Kashmir State on your election as Sadar-i-Riyasat. I should like to congratulate the people of the State also on their wise choice. This puts a great responsibility upon you, for you have not merely to follow an established convention but rather to help in making conventions for the future. You know how dear the future of the State is to me. It is dear to me because of my own intimate relationship with Kashmir and it is dear to me also because of the numerous ties that bind the State to India. Our future is linked together and we have to face good fortune and ill-fortune alike together.

A new chapter opens now in the Jammu and Kashmir State. And yet, although it is new, it is a continuation of the old but in a different form. The processes of life, whether that of an individual or of a nation, are both a continuation and a continuous change.

I earnestly hope that the changes that have been brought about in the Constitution of the Jammu and Kashmir State will lead to the greater prosperity and happiness of the people of the State and will bring them even closer to India, of which they are such an intimate part.

To you, who have to shoulder this burden and this responsibility at such an early age, I send all my good wishes and my affection.

Yours sincerely,
Jawaharlal Nehru

Characteristically, he appended a small note:

My dear Tiger,

I am sending you separately a semi-formal letter of congratulation and good wishes. I did not quite know how to address such a letter. You know that I shall often think of you and that you can always rely on such help and guidance as I can give.

Yours
Jawaharlal Nehru

On the night of the 16th Asha and I sat up into the small hours of the morning. I knew that I had irrevocably severed my links with the feudal system and also that, howsoever cordial our relations would be on the surface, my father would not easily forgive me for accepting the new office. I realized that the Sheikh was an inveterate foe of our dynasty and that by accepting it I was virtually placing myself at his tender mercies. I also knew that the reaction among my own community, the Dogras of Jammu, would be hostile, at least to begin with. And yet I was convinced that the old order had passed never to return; that whatever the future might hold there was no future for me or my people unless I threw in my lot with Jawaharlal Nehru and the new India that he had done so much to create and was guiding with such courage and foresight. While I was cutting myself away from the old feudal tradition, I felt that by accepting the new challenge I was also joining the much larger adventure of building and shaping the destiny of a nation representing one-seventh of the human race. The die was cast, and on the morrow I would move onto a fresh stage of my life and destiny.

The next morning, 17 November 1952, I drove to the old Rajgarh Palace on the Jhelum which had been the residence of the former rulers until my father shifted to the Dal Lake. The Darbar Hall with its magnificent papier mâché ceiling, the symbol *par excellence* of the old order, had been converted into a legislative chamber for the Constituent Assembly. Where previously the courtiers had sat on the carpet with only my father sitting on his gold throne, now benches had been constructed and the Speaker sat on an elevated platform at the opposite end of the beautiful hall. It was to this platform that I was conducted, having been received at the entrance by Sheikh Abdullah, Bakshi Ghulam Mohammad and the Speaker, G. M. Sadiq. As I entered the hall the assembled members of the Constituent Assembly stood and applauded vigorously. I was at once the last representative of the old order becoming, by the consent of the people, the first servant of the new.

After I took my seat on the dais, flanked by the Sheikh and his colleagues, the Chief Justice Wazir Janki Nath stood up and administered the oath of office which I repeated after him. Then I read out a short speech that I had worked on for several days in which, *inter alia*, I said:

> I am aware that the functions of the Head of a State are, as a focus of unity, both important and onerous. This is particularly true in a State where conditions have been as abnormal as they have been here. I must admit that I had some hesitation in taking upon myself the responsibilities which this position entails, knowing as I did that there must be many better fitted than myself both in ability and experience to fill this office. But the trust and confidence which has now been placed in me has given me hope and courage to assume these new responsibilities, and I can assure you that whatever talent and capacity I might possess will be entirely at the service of the State and its people.
>
> Our State has, as I have said, been passing through extremely abnormal times, and over the last few years it has been subjected to extraordinary strains and stresses. The ruthless and cruel invasion of our fair country brought in its train untold suffering and misery. Hundreds were massacred, thousands more were uprooted from their hearths and homes and thrown into a nightmare of horror and wretchedness. In the long and chequered history of our land this was perhaps the greatest crisis that our people were called upon to face and combat. It is a tribute to them and to their leaders that they rose to the occasion and met the situation with composure and courage. This heroic endeavour alone would not have availed, had it not been firstly for the gallant resistance put up by our forces, isolated and overwhelmingly outnumbered as they were, and secondly for the

timely aid which was rushed to us by India in that hour of greatest need.

I concluded with the following words:

> Our State can be in a position to face all these vital issues only with the united strength of our people. In this land of colour and beauty, men of different faiths and creeds live as the common inheritors of a great past and culture. It is our task now to forge a greater unity among them as the joint architects of their future. Such abiding unity cannot be imposed from above but has to be based upon the interests of the common man in all parts of the State. In building this equal partnership of all the people and all the regions of this State, a solemn duty devolves upon each one of us to do our bit, to make our individual contribution. With your blessings and good wishes I hope to be able to effectively contribute towards this end.

Chapter 12

A week after assuming the office of Sadar-i-Riyasat the Government moved down to Jammu for the winter. The Praja Parishad had of course been outraged at my acceptance of the office, and had threatened a black-flag demonstration on the day I arrived in Jammu. On 24 November I flew down on the Indian Airlines plane. In sharp and poignant contrast to the earlier occasions after my return from America and my marriage, when the people of Jammu gave me an enthusiastic and affectionate welcome, this time there were derisive and hostile slogans, and the city, from the airport right up to the palace gates, was a sea of black flags. Bakshi Ghulam Mohammad was with me in the open jeep, and though the National Conference had tried to lay on some kind of reception it was swamped by the deep hostility of the Dogra masses.

I must admit that it was a rather traumatic experience, but I put up a brave front and continued to smile and greet the people. I noticed that, despite themselves, many waved back. The demonstration, in fact, was not so much against me as a gesture of loyalty to and solidarity with my father. It reflected the widespread agitation launched by the Praja Parishad on 14 November against Sheikh Abdullah. Their slogan of complete integration of the State with India was expressed in the rallying cry '*Ek vidhan, ek nishan, ek pradhan*' (one constitution, one flag and one president). This agitation gathered momentum over the next few months, as it effectively capitalized upon the sense of outrage felt by the Dogras not only at having lost their predominant position in the State but also in having at one stroke been placed at the mercy of their arch enemy, Sheikh Abdullah. The Sheikh, for his part, not only made no effort to mollify the feelings of the Jammu people but continued with his hostile and aggressive attitude. An example was the question of the flag to be flown on the Jammu Secretariat. The old State flag having been hauled down, I had suggested to the Government that, along with the new flag, the national flag should also be hoisted. This was sharply turned down by the Sheikh, and so

I in turn declined the suggestion that I should personally hoist the new flag.

After settling down in Jammu I began assessing the situation, and soon realized that the Praja Parishad agitation had spread deep and wide throughout the Jammu region. Even though I knew Jawaharlal's aversion to the Parishad, I felt it my duty to inform him about the true state of affairs. I prepared a detailed note which analysed the whole position and made certain concrete recommendations for political and economic measures that would help to meet the genuine aspirations of the people of Jammu and Ladakh. As I put it, 'Stripped of all its non-essentials, the situation is that whereas Jammu and Ladakh strongly desire complete integration with India, Sheikh Sahib and his colleagues are extremely insistent upon the "limited" nature of the Accession and are not prepared to agree to complete integration.' This I sent to Jawaharlal with a covering letter on 22 December, following up an earlier one dated 1 December 1952. I was quite frank. 'The situation is serious', I wrote, 'not in any military sense but in the sense that an overwhelming majority of the Jammu province seem to be emphatically in sympathy with the agitation. Fundamentally responsible for this, I feel, are several deep-seated and genuine economic and psychological reasons, and I do not think it will be a correct appraisal to dismiss the whole affair as merely the creation of a reactionary clique.'

I followed this with a visit to Delhi where I talked with Jawaharlal, the Home Minister, Dr Katju and Dr Rajendra Prasad, and gave them my assessment of the situation. I urged that the Government of India must intervene so that the State Government reacts to the Jammu agitation not only with police repression but with concrete political, economic and administrative measures. Jawaharlal did write to Sheikh Abdullah in the same month, who replied with a long letter justifying the stand of his Government and rejecting the whole Jammu problem as the work of communal organizations and a 'violent reaction on the part of Jammu landlords and other upper classes'. Jawaharlal also answered my letter, saying, 'I can quite understand your great concern over the developments in Jammu. I am naturally also very much concerned and I have followed them closely. I entirely agree with you that while police measures are of course necessary, that is only a negative way of dealing with the situation. . . . The situation in Jammu is serious enough to deserve our fullest consideration and such positive action as may be necessary. At the same time, one has to view these matters, as all other important matters, coolly and dispassionately.'

Later that month I received a message from Jawaharlal saying that the President had invited me to attend the Governors and Rajpramukhs Conference being held early in February and that he would like me to attend. This was my first opportunity to attend a meeting on the national level, all my previous experience having been connected only with Jammu and Kashmir. It was, therefore, with some excitement that I accepted the invitation. I arrived in Delhi on 3 February and was put up at Rashtrapati Bhavan in the huge Dwarika Suite reserved for visiting dignitaries. After a formal call on Dr Rajendra Prasad I turned in and went to sleep quite early. I had a curiously vivid dream that night. I was standing in a large room, and Mahatma Gandhi came in. I remember him distinctly, all his features and clothes much clearer in my memory than when I had actually seen him in Srinagar several years earlier. He walked up to me, put his left hand on my shoulder and took my right palm in his other hand. He looked at it for a moment, and then said in English, 'You will be a very wise man'.

I found the two-day Conference itself interesting, the first of a long series of fifteen such consecutive conferences that I was to attend in the future. It opened with a formal speech read out by the President, after which Jawaharlal made an hour-long intervention with a wide ranging survey of the current national and international scene, laying special emphasis upon various aspects of economic development. The scholarly Vice-President, Dr Radhakrishnan, spoke on the second day, brilliantly outlining his views on education and national development. In those days the princely States had yet to be fully integrated into the republic, and were grouped into several geographical zones each headed by a senior prince called the Rajpramukh. The Conference was thus attended by the former rulers of Mysore, Nawanagar, Patiala, Travancore, Gwalior and Jaipur, apart from the Governors who included Chandulal Trivedi, K. M.

Munshi, R. R. Diwakar, Jairamdas Daulatram, Fazl Ali and Pattabhi Sitaramaiyya.

At twenty-two I was, of course, one-third the age of most of the participants. After the general speeches each participant gave a brief report of the situation in his State, highlighting the main problems that they were facing. In my remarks I reviewed the important developments over the last year, and also mentioned the continuing agitation in Jammu. Apart from a formal banquet hosted by the President and attended by Cabinet Ministers, Jawaharlal held a dinner for the participants. He also had me over for a small, private lunch at which Indira Gandhi was hostess and Padmaja Naidu the soul of the party.

Meanwhile the Jammu agitation showed no signs of abating and, in fact, was taken up by like-minded parties in Delhi, specially the newly formed Bharatiya Jan Sangh headed by Dr Shyama Prasad Mukherjee and N. C. Chatterjee, who decided to launch a Satyagraha on an all-India basis. I continued to urge that there should be a dialogue between the State Government and the agitation leaders, but Sheikh Abdullah would have none of it and Jawaharlal was also averse to such talks. As he wrote in one of his letters to me (22 March 1953): 'In my view what these people have done is little short of treason to the country and the people should realize it.' My suggestion to Sheikh Abdullah that my formal address to the Legislative Assembly on 25 March could be used as a good opportunity to strike a conciliating note was rejected by him.

Although Sheikh Abdullah attempted to dismiss the whole matter as the plot of 'reactionary elements', and the Government of India seemed in the beginning to support that view, I was deeply disturbed because I realized that the final chance of building a new Dogra–Kashmiri rapport, which alone could ensure the stability and welfare of the State, was being lost. I continued to dwell on this theme in my letters to Jawaharlal, even though I knew that he was allergic to the Jan Sangh and Praja Parishad leaders. In a letter of 27 March I wrote: 'What really disturbs me is the fact that the gulf between Jammu and Kashmir has widened tremendously over the last few months, and that the breach instead of being bridged seems to be steadily widening. Neither of the parties seems to quite realize the implications of this, and I fear that we may reap a very bitter harvest in years to come.'

At about this time Bakshi Ghulam Mohammad, who was Deputy Prime Minister, and Girdhari Lal Dogra, the Finance Minister and sole representative in the Cabinet from Jammu, met Jawaharlal in Delhi to discuss the whole situation. Bakshi was very different to the Sheikh, more pragmatic, a superb organizer and a man with excellent public relations with all sections of pecple including many in Jammu. Although closely associated with Sheikh Abdullah and the National Conference, he never displayed the aggressive anti-Dogra attitudes of the Sheikh and M. A. Beg. His whole stance *vis-à-vis* the Accession was also distinctly more amenable to strengthening the relationship between the State and the Centre, and less charged with the Kashmiri chauvinism so sedulously fostered by Sheikh Abdullah.

Apart from the Jammu agitation, there was the broader question of implementing the 'Delhi Agreement' on the constitutional relationship between the State and the Centre that had been earlier worked out after prolonged negotiations between Sheikh Abdullah's team and representatives of the Government of India. While the decision regarding abolishing the monarchy had been rapidly implemented because it suited the Sheikh, he had begun dragging his feet as far as the other items were concerned. It had been clear to me from the start that he was less than sincere in his professions, and that he was simply manoeuvring himself into a position of strength in relation to the Central Government. Had he been sincere, why the delay in implementing the rest of the agreement?

In fact the Sheikh used one of his favourite techniques by initiating a debate within the National Conference Working Committee regarding the nature of the relationship with India. One group, led by M. A. Beg, was vociferous in its view that the relationship should not go beyond the three subjects contained in the Instrument of Accession, while the other, led by Bakshi Ghulam Mohammad and including G. L. Dogra and D. P. Dhar, was agreeable to a more comprehensive relationship covering other vital areas such as the judiciary, financial arrangements, and so on. This tussle within the National Conference gradually assumed serious dimensions. While the Sheikh ostensibly tried to keep himself above the controversy for some time, he increasingly began throwing his weight in favour of the hardliners. It was this development, even more than the Jammu agitation, that began to disturb Jawaharlal.

I had two long meetings with him on 21 April and 23 May. At the second meeting he opened up for the first time, and expressed his deep anguish at the way the situation in the State was developing. He admitted that he had no answer when asked in Parliament or outside as to why the Delhi Agreement had not been implemented. He said that he had written a long letter to the Sheikh pointing out that the matter had been greatly delayed and that, as he was going abroad in a few weeks, he would very much like to see the question finalized before he left. When I asked him what reply he had received he turned to me and said in an incredulous and hurt tone, 'I received no reply at all'. He added that the Sheikh was thoroughly confused and was evidently avoiding meeting him.

I mentioned that if the State Government made any move which showed that they were going back on their solemn agreements with the Government of India, my position would become impossible. It would be very difficult for me to continue to associate myself with them as I would never be a party to a betrayal of India. He pondered over that for a while but did not immediately respond. When I got up to leave he turned to me and said, 'Look here, one point I quite agree on is that if our agreements are thrown overboard or something like that, your position becomes absolutely impossible.' It was clear that he was deeply disturbed, but he had not quite made up his mind how to deal with the problem. He was particularly hurt, even bewildered, at the hostile manner in which his old protégé and friend Sheikh Abdullah was acting, and agreed with me when I remarked that unlimited power seemed to have brought out the Sheikh's worst fascist and totalitarian tendencies.

Indeed, the Sheikh's attitude steadily became more and more intransigent. He made a speech at Ranbirsinghpura, a border town near Jammu, where he reacted violently against the Jammu agitation, went on to accuse India of being communal and virtually threatening that the Accession of the State could not be taken for granted. I had been planning a visit to Europe, and had even written formally to the President and Jawaharlal about this, but in view of the growing tension following the Jammu agitation and the increasing schism within the National Conference I decided against going.

The annual move to Srinagar saw further deterioration. Dr Shyama Prasad Mukherjee, who had defied the ban on his entry to the State, had been arrested and kept in detention. On 10 June I sent a report to Jawaharlal, who was at that time on a fairly long visit to London, in which I said:

> The political situation here in the valley continues to be extremely fluid. The division within the party is causing considerable tension. The pro-Indian faction continues to be determined, and claims to be strong and to have a majority both in the Working Committee and in the Assembly. Frequent meetings of the Working Committee continue.
>
> I was shocked and astounded to gather from a private meeting with Sheikh Abdullah last week that he seems to have decided to go back upon the solemn agreements which he has concluded with India and upon his clear commitments. This cannot be allowed, as it will make our position absolutely impossible and be a grave blow to our National interests and naturally to our International position also. I need not mention the grave and widespread repercussions that will result from such a development. The problem will claim your immediate attention upon your return for a final and decisive solution.

Soon thereafter came the shocking news of the death of Dr Shyama Prasad Mukherjee in detention. I was not informed of his illness or his removal to hospital, and only learnt of his death from unofficial sources several hours after his body had been flown out of Srinagar. The circumstances in which he died in the custody of the State Government were a cause of grave resentment and suspicion. Jammu was furious because Dr Mukherjee had been martyred while fighting for the Praja Parishad cause, and there was open talk that his death had not been from natural causes. The whole of India was shocked at this event, specially the people of Bengal, who held Dr Mukherjee in the highest regard.

By now the Sheikh was clearly on the warpath. Despite repeated suggestions to visit Delhi, and an invitation from Jawaharlal to do so on 3 July, he refused to go down to the capital and discuss the whole situation. Maulana Azad came up to Srinagar for a few days, but instead of taking advantage of his visit to sort out differences, Sheikh Abdullah pointedly ignored him and he was virtually insulted by the National Conference workers. The Sheikh's speeches became more and more strident, and it become increasingly clear that he was seriously working on the idea of some sort of independent status for Kashmir which, inevitably, would imply a virtual negation of the Accession to India. At about this time Adlai Stevenson visited Srinagar and had long talks with the Sheikh. It is not known what exactly transpired, but the general impression was that in some way the Sheikh received encouragement from these conversations for his independence theory.

Meanwhile, the rift within the National Conference had come into the open. It was common knowledge that the Sheikh was favouring the hardliners led by M. A. Beg, while most of the other senior leaders, including the two Cabinet Ministers G. L. Dogra and Shamlal Saraf, and D. P. Dhar who was Deputy Minister, had rallied behind Bakshi Ghulam Mohammad. Bakshi came to my house on a few occasions in connection with meetings of the State Soldiers Board of which I was Chairman, and would stay on for a few minutes after the others had left. While I was careful not to give the impression of getting involved in the controversy, I had of course to maintain close contact with the pro-Indian group. D. P. Dhar was a frequent visitor, and he became one of the key figures in the drama that was about to unfold. Suave and unflappable, 'D.P.' had an incisive mind and was an excellent planner. He played a significant role in keeping New Delhi informed about the inner conflicts within the National Conference. Jawaharlal liked him and had regard for his political judgement.

It was at about this time that we began to realize that unless something drastic was done to curb Sheikh Abdullah, the situation would steadily deteriorate and finally get completely out of hand with grave and incalculable consequences for the entire country. The Kashmir issue was still a prominent item on the Security Council agenda, and it would be disastrous if the Sheikh, who had twice been sent to Lake Success as a member of the Indian delegation, were to do a volte-face while still Prime Minister of the State. I watched with increasing alarm and apprehension the trend of events, and decided that I should once more go down to Delhi and discuss the situation with Jawaharlal. This I did in the third week of July.

When I met him, I found Jawaharlal's attitude considerably changed. Not only did he not make any attempt to defend Sheikh Abdullah, he seemed to be as disturbed as I was about the way the situation was developing. It seems that, apart from my own letters, he had received detailed reports from the Intelligence Bureau (then headed by B. N. Mullick), D. P. Dhar and others from the State, as well as first-hand impressions from Maulana Azad and Jawaharlal's close political confidant, Rafi Ahmed Kidwai. He listened in grim silence to my detailed presentation, occasionally frowning and nodding agreement. I did not put any concrete proposal before him, but I did make it clear that if Sheikh Abdullah persisted in his hostile attitude a parting of the ways was inevitable. When I left he got up and saw me to the door. As I took leave he put his hand on my shoulder and said, 'Don't worry, do your best.'

Immediately upon my return to Srinagar I decided to break completely with the problem on hand and go off on a pilgrimage to Amarnath, the famous shrine dedicated to Lord Shiva and situated in a huge cave at an altitude of 13,000 feet. I had always wanted to go there, and felt that this was an appropriate time to do so. It would give me, hopefully, fresh strength to deal with the looming crisis. Also, as the trip would be well publicized, it would allay any suspicions that may have been raised by my airdash to Delhi. The main pilgrimage to Amarnath reaches the shrine on the full moon of Shravana, which falls sometime in August. However, some devotees go a month earlier, and the July full moon was on the 26th. We therefore left Srinagar on the 23rd.

The trip was one of the most memorable I have ever made. From Pahalgam I was carried in a *dandy*, as my leg was still not good enough to manage the steep climbs, while Asha insisted on walking all the way. We camped for three nights—at Chandanwari, Sheshnag and Panchtarni—before reaching the famous cave. The scenery *en route* was breathtaking, specially the fabulous milky-green Sheshnag lake set against a massive, triple-peaked glacier. Once again I was struck by the uplifting power of natural beauty, particularly in the higher altitudes. As I wrote in a series of articles published by the *Hindustan Times* in its Sunday supplements:

One distinctly feels the presence of a Power greater, stronger and purer than one's own petty self. For a while I glimpse the lovely face of nature in its pure undesecrated majesty. The torrent of time slackens, the problems and strifes of life pale into insignificance and I am lost in deep contemplation. I cherish a desire in the recesses of my heart to one day build myself a small Ashrama in such surroundings where, with the body and mind made pure and free from the tentacles of desire and fear, ego and attachment, one can concentrate upon the unalloyed purity of Nature and thereby perhaps achieve spiritual illumination.

These articles were subsequently published in a small booklet called *The Glory of Amarnath*, which was my first literary endeavour.

The cave itself was much bigger than I had imagined, and in one corner stood the glistening ice formation about five feet high, symbolizing the creative energy of Lord Shiva. The shrine is unique in that the ice *lingam* forms of itself every year, and is believed to wax and wane with the moon. On the journey I was reading Paul Brunton's *Search in Secret India*, and the whole experience made a deep impression upon me. I have found repeatedly that an outer crisis is an excellent time to deepen one's inner aspiration, that the thicker the battle presses the more significant becomes the voice of the inner charioteer.

I returned to Srinagar on 28 July. During my absence there had been a fresh series of meetings of the National Conference Working Committee at which the two sides had clashed—Sheikh Abdullah was now openly partisan in the controversy, but only one other member of his Cabinet, M. A. Beg, supported him. The rift in the Cabinet came to a head on 7 August when the Sheikh, using a flimsy pretext, decided to move against the opposing group by asking for the resignation of Pandit Sham Lal Saraf. On the 8th morning Saraf sent me a copy of a long letter to the Sheikh in which he accused him of having repudiated the declared policies of the National Conference concerning the relationship of the State with the Indian Union. He refused to resign saying that 'the manner in which you have created a dangerous situation in the country by making highly inflammable speeches before the public, combined with your authoritarian attitude in the Cabinet, have convinced me that instead of helping the difficult situation my resignation will encourage you to pursue your policies unbridled. Such a course will be suicidal for the country.'

A few hours later I received a signed copy of a memorandum sent to Sheikh Abdullah by the Deputy Prime Minister, Bakshi Ghulam Mohammad, the Finance Minister, G. L. Dogra and the Health Minister, Pandit Sham Lal Saraf. The five-page document squarely accused the Sheikh and M. A. Beg of blatantly flouting accepted party policies. It said:

After convening of the Constituent Assembly, certain inescapable elaborations of the State's relationship with India were defined in the Delhi Agreement, of which you were the Chief architect on our behalf. Your stand was unanimously endorsed by the Government, the National Conference, the Indian Parliament and the Constituent Assembly of the State. But you have not only deliberately delayed the implementation of the Agreements on these matters, which form the sheet-anchor of our policy, but have purposefully and openly denounced these in public. You have thus arbitrarily sought to precipitate a rupture in the relationship of the State with India.

They added:

Mr M. A. Beg has persistently been following policies of narrow sectarianism and communalism, which have seriously undermined the oneness of the State. Unfortunately, you have been lending your support to his policies in the Cabinet and his activities in public. This has generated bitter feelings of suspicion and doubt in the minds of the people of the various constituent units of the State. You have connived at all these unfortunate happenings and thus strengthened and encouraged the forces of disruption. The result is that unity and secular character, the two fundamental aspects of our State, stand threatened today.

Concluding, the three Ministers said:

We have been constantly urging upon you to put an end to these unhealthy tendencies and to undertake unitedly measures for restoring the morale of the people. In spite of our best intentions, we have failed in our efforts. It is, therefore, with great pain that we have to inform you of our conclusion that the Cabinet, constituted as it is at present and lacking as it does the unity of purpose and action, has lost the confidence of the people in its ability to give them a clean, efficient and healthy administration.

As soon as I received this communication, which was not entirely unexpected, the ball was squarely in my court. Legally the Council of Ministers held office at the pleasure of the Sadar-i-Riyasat, and as the appointing authority I also enjoyed the power of dismissal, even if this was not specifically spelt out. However, before taking any drastic action, I thought it was only appropriate that I should talk with Sheikh Abdullah. I immediately invited him to come across to see me as soon as possible. He was going to Gulmarg that afternoon, and came to my residence around noon. When I inquired about the situation he narrated three long though petty incidents regarding Shamlal Saraf, and disclosed that he had asked him to resign. He mentioned that the Kashmir Trade Commissioner in Delhi had telephoned him in the morning and informed him that the newspapers carried big headlines about the 'constitutional crisis' in Kashmir, and expressed surprise at the leak.

I said that I was deeply distressed and concerned at the recent trend of events, particularly the complete absence of homogeneity in the Cabinet. I pointed out that it would be useful if he and his Cabinet colleagues were to come to my residence that evening so that the whole matter could be discussed in depth. He sidetracked this suggestion by launching into an angry tirade against the Indian Press which he accused of grossly misrepresenting and exaggerating the differences in the Cabinet. He made the astonishing claim that, although there was some difference of opinion, there were no basic political or administrative differences within his Cabinet. When I again stressed the necessity of doing something to rectify the situation he stated quite clearly that no internal solution was possible until there was some external solution acceptable to India as well as Pakistan.

This revealed his true state of mind. From a position of clearly endorsing the Accession to India as 'final and irrevocable', he had over the last few months moved into an entirely different posture. He was evidently expecting some international pressure to be brought upon India to grant virtual independent status to Kashmir, and this was quite obviously the reason why he was holding up the implementation of the Delhi Agreement which would further cement the relationship between the State and the Centre. In our meeting he was not only unable to give me any assurance that the internal crisis in his Cabinet would be resolved, but had added a new dimension of uncertainty by talking about an 'external solution'.

Our meeting lasted for about forty-five minutes, after which he left saying that he was going to Gulmarg for the weekend. It was quite clear to me that something had to be done at once if further drift was to be avoided. I summoned my political and legal advisers, including D. P. Dhar and Brigadier (later General) B. M. Kaul, who was acting as a sort of unofficial courier between us and Delhi. The Sheikh was still a popular figure in the valley despite the fact that corruption and maladministration had to some extent eroded his stature. If we gave him an opportunity to take his case to the streets he would easily arouse acute communal and chauvinistic sentiments among the Kashmiri masses, which could in turn result in serious and violent disturbances. Anti-national elements and agents were active in the valley and, if given the chance, would not hesitate to plunge the State into utter turmoil. Parliament was in session, and anything happening here would at once have its repercussions on the national scene.

We, therefore, decided that the Sheikh would have to be dismissed. While I was reluctant that he should be arrested, Bakshi Ghulam Mohammad made it quite clear that he could not undertake to run the Government if the Sheikh and Beg were left free to propagate their views. Dismissing and detaining a sitting Prime Minister was no easy task, as all plans had to be made in absolute secrecy. The police had to be organized, the Indian Army kept discreetly informed, and the media handled in an imaginative manner. Luckily, D. P. Dhar was a superb tactician who was at his best in such situations. In the three days preceding the crisis 'D.P.', who was Deputy Home Minister, had tied up all these details.

The fact that the Sheikh decided to go up to Gulmarg, however, necessitated some last minute change of plans. Soon after he left my house we sat down to draft the dismissal order. Along with this I dictated a covering letter to Sheikh Abdullah. The text of these documents was as follows:

SECRET

Karan Mahal,
Srinagar
August 8, 1953.

My dear Sheikh Abdullah,

You will recall that in the course of our meeting today I conveyed to you my deep concern at the serious differences which exist in your

Cabinet. I impressed upon you the immediate necessity for restoring harmony and unity of purpose among the members of the Cabinet in the execution of its policies. You were, however, unable to assure me that these acute differences could be remedied.

This conflict within the Cabinet has for a considerable time been causing great confusion and apprehension in the minds of the people of the State. The situation has reached an unprecedented crisis with the fact that three of your four Cabinet colleagues have in a Memorandum to you, a copy of which they have sent to me, expressed their complete dissatisfaction with your actions and policies, which have lost the present Cabinet the confidence of the people. This document clearly indicates that the divergence within your Cabinet has reached proportions in which the unity, prosperity and stability of the State are gravely jeopardized.

When we met today I further suggested to you that an emergent meeting of the Cabinet should be held at my residence this evening so that we could jointly explore the possibilities of securing a stable, unified and efficient Government for the country. But to my regret you evaded the issue.

Under these conditions I, as Head of the State, have been forced to the conclusion that the present Cabinet cannot continue in office any longer and hence I regret to inform you that I have dissolved the Council of Ministers headed by you. A copy of my order in this connection is attached herewith.

I need hardly add how deeply distressed I am at having to take this action, but the vital interests of the people of the State, which it is my duty to safeguard, leave me no alternative. I trust that this will in no way affect the mutual regard and cordial feelings we have for each other.

Yours sincerely,
Karan Singh
SADAR-I-RIYASAT

ORDER

Whereas for some months I have been noticing with growing concern that there have existed acute differences of opinion between members of the Government on basic issues—political, economic and administrative—affecting the vital interests of the State;

And whereas members of the Government have been publicly expressing sharply conflicting points of view regarding these matters;

And whereas on these fundamental issues the views of a majority of the members of the Cabinet are sharply opposed to the view held by the Prime Minister and one of his colleagues;

And whereas efforts to work in harmony and pull together as a team having failed, and the majority in the Cabinet has expressed that, lacking as it does in unity of purpose and action, the present Cabinet has lost the confidence of the people;

And whereas the economic distress of the people has considerably increased which needs prompt and serious attention;

And whereas a stage has reached in which the very process of honest and efficient administration has become impracticable;

And whereas, finally, the functioning of the present Cabinet on the basis of joint responsibility has become impossible and the resultant conflicts have gravely jeopardized the unity, prosperity and stability of the State;

I, Karan Singh, Sadar-i-Riyasat, functioning in the interests of the people of the State who have reposed the responsibility and authority of the Headship of the State in me, do hereby dismiss Sheikh Mohammad Abdullah from the Prime Ministership of the State of Jammu and Kashmir, and consequently the Council of Ministers headed by him is dissolved forthwith.

SADAR-I-RIYASAT
JAMMU & KASHMIR

Srinagar,
August 8, 1953

By the time the documents were ready it was late evening. The weather was also in a turbulent mood; it was pouring, thunder rumbled and jagged shafts of lightning cut through the clouds. I deputed my ADC, Major B. S. Bajwa, to go to Gulmarg and deliver the letter to Sheikh Abdullah. A police party also went along with him, but they were delayed by the torrential rain and the bad road beyond Tangmarg. Having dispatched the documents we all waited in Karan Mahal for word that they had been delivered. As this was delayed, tension grew. Our gamble was a risky one, for if the Sheikh got even an inkling of what was happening he would react ferociously, and our own lives may well have been in danger. However, the die was cast, and all we could do was to pray that the whole operation went off smoothly.

As it turned out, our prayers were answered. The Sheikh was totally unaware of the developments, and so arrogant in the possession of power that he could never dream that anyone would dare to challenge him. When the ADC and the police contingent finally reached Gulmarg it was late at night, and the

Sheikh and Begum Abdullah were fast asleep. With some difficulty and much knocking he was awoken and handed the letter as well as a warrant of arrest. On reading it he flew into a rage and shouted, 'Who is the Sadar-i-Riyasat to dismiss me? *I* made that chit of a boy Sadar-i-Riyasat.' But by then his house was fully surrounded by the police. He was given two hours to say his *namaz* and pack, during which we later learnt he burnt a number of documents that he had with him. This could have been prevented, but we had given strict instructions to the police that he and the Begum were to be treated with courtesy and not physically harassed in any way. In the early hours of the 9th morning they were escorted to a car and driven out of the valley to the Tara Niwas guest house in Udhampur, where they were kept in detention. M. A. Beg and several others were also arrested that night in Srinagar and other parts of the valley.

Meanwhile, it was my responsibility to see that there was no constitutional gap in the administration of the State. With the dismissal of Sheikh Abdullah I wrote to Bakshi Ghulam Mohammad inviting him to meet me and discuss the formation of the new Government. I ended the letter by saying that 'the continuance in office of the new Cabinet will depend upon its securing a vote of confidence from the Legislative Assembly during its coming session'. Bakshi came over and we discussed the situation. I felt that no time should be lost in swearing in the new Government. We summoned the Chief Secretary, M. K. Kidwai, who was blissfully unaware of what was happening. When he learnt that the Sheikh had been dismissed and arrested he sank down on the steps holding his head in his hands, and was revived only after he had been plied with a couple of stiff whiskies.

In the early hours of 9 August I administered the oath of office to Bakshi Ghulam Mohammad and G. L. Dogra, so that they would be in a position to deal with the tense situation that would inevitably develop as soon as the news of the dismissal and arrests spread in the valley. I then wrote a report to the President informing him of the entire developments, and sent a copy to Jawaharlal with a covering letter which included these two paragraphs:

We are all acutely aware of the gravity of the step that we have been forced to take and also of the extent of its possible repercussions, both within and without the State. In this whole matter I have attempted to act in a democratic and constitutional manner, keeping especially in mind what you said when we met last. On the whole I feel that we have done the best that was possible under the circumstances.

Regarding what was to be done with Sheikh Abdullah after his dismissal, this was of course a decision for the new Government to take. On my part I strongly urged them to desist from arresting him soon after the dismissal, but they were most apprehensive that with his presence in the valley at this juncture reactions would have been greatly intensified and there was a grave danger of the situation getting completely out of hand and even resulting in violence and bloodshed. Consequently he has been arrested this morning at Gulmarg and is being taken to Udhampur where he will be lodged in the State Guest House. I have stressed that he and his family be accorded all courtesy and consideration.

By the time the whole operation was over dawn had broken. The clouds and thunder of the last two days had disappeared, and the sky was clear again. I got through to Jawaharlal on the telephone just before eight but the line was unclear. In its morning bulletin All India Radio devoted thirteen of its fifteen minutes to a detailed report on the developments in Kashmir. I had not slept a wink that night and found myself in a curiously detached and light-headed mood. I knew that what I had done was for the national good, and that all the danger and risks were worth taking if it served the country. Asha had also sat up anxiously in her room all night. Slowly I climbed up the steps and walked into her room. 'It is over', I said, and we both smiled; she sixteen going on seventeen, I twenty-two going on twenty-three. I was asleep almost before I hit the bed.

Sadar–i–Riyasat

Chapter 13

The dramatic developments in Jammu and Kashmir, culminating in my dismissal of Sheikh Abdullah and installation of Bakshi Ghulam Mohammad as Prime Minister of the State in his place, evoked a sharp, almost hysterical reaction in Pakistan. Furious at what they considered to be a 'pro-Indian' coup, they let loose a barrage of vituperative propaganda against the new regime headed by Bakshi Ghulam Mohammad. The press and radio, especially the so-called 'Azad Kashmir Radio', ran extremely provocative and inflammatory anti-Hindu and anti-India broadcasts aimed at exploiting religious fanaticism and inciting the people of the valley to revolt against the new government. At the same time their already fairly extensive espionage activities within the State were stepped up.

In April 1953 Prime Minister Nazimuddin of Pakistan had been dismissed by the Governor-General, and Mohammad Ali appointed in his place. At the Commonwealth Prime Ministers' Conference in London in June he and Jawaharlal Nehru had met and discussed various matters including Kashmir and, on 25 July, only a few weeks before Sheikh Abdullah's dismissal, Jawaharlal had paid a visit to Karachi, where he received a tumultuous welcome. Talks were held in a cordial atmosphere, with both Prime Ministers saying that although no solution to the Kashmir problem had been found, they had secured a much better understanding of each other's position and that the next round of talks would be held in Delhi towards the end of August.

As soon as the news of Sheikh Abdullah's dismissal broke, the Pakistan government reacted strongly. Mohammad Ali sent a message to Jawaharlal suggesting an immediate meeting, and despite the latter's marked lack of enthusiasm he finally agreed to bring forward the talks. Mohammad Ali arrived in Delhi on 16 August and talks began, concluding on the 20th with a joint communiqué. This, *inter alia*, stated:

Both the Prime Ministers were actuated by a firm resolve to settle these problems as early as possible peacefully and co-operatively to the mutual advantage of both countries. The Kashmir dispute was especially discussed at some length. It was their firm opinion that this should be settled in accordance with the wishes of the people of that State, with a view to promoting their well-being and causing the least disturbance to the life of the people of the State. The most feasible method of ascertaining the wishes of the people was by a fair and impartial plebiscite. Such a plebiscite had been proposed and agreed to some years ago. Progress, however, could not be made because of lack of agreement in regard to certain preliminary issues. The Prime Ministers agreed that these preliminary issues should be considered by them directly in order to arrive at agreements in regard to them. These agreements would have to be given effect to and the next step would be the appointment of a Plebiscite Administrator. In order to fix some kind of a provisional timetable, it was decided that the Plebiscite Administrator should be appointed by the end of April 1954. Previous to that date, the preliminary issues referred to above should be decided and action in implementation thereof should be taken.

This news was on the radio that evening, and evoked a sharp reaction from Bakshi. He dashed off a highly emotional letter of resignation in his own handwriting and sent it to my residence. For obvious reasons I did not take any action on it, and immediately telephoned Delhi. Jawaharlal had deputed Ajit Prasad Jain and Vishnu Sahay to Srinagar to explain the position to Bakshi. On the 20th evening all of us met at Karan Mahal, with Bakshi still in a state of shock and sulking at his residence. When I informed Jawaharlal on the telephone of Bakshi's resignation he lost his temper and spoke angrily of this 'ridiculous nonsense'. I handed the phone to Ajit Prasad who listened for a while until Jawaharlal banged down the phone. We then went to work on Bakshi, and finally persuaded him not to press his resignation. The whole incident was kept a closely guarded secret, because any hint that Bakshi was wavering for whatever reason would have been disastrous at that critical juncture when the law and order situation was just beginning to come under control.

What had really upset Bakshi was the deadline for the appointment of the Plebiscite Administrator. He felt, with some justification, that if India was really going ahead with the plebiscite his helping to oust Sheikh Abdullah had placed him in an absolutely false position *vis-à-vis* the people of the State.

However, when we explained the position to him he relented, and on 21 January issued a statement: 'This declaration was finalized on behalf of the Government of India with our concurrence and has our unqualified support.' What might have been a serious crisis blew over. As it turned out, the subsequent correspondence between Jawaharlal and Mohammad Ali showed a major difference in perceptions, and the decision of the USA to give military assistance to Pakistan, which became apparent towards the end of the year, introduced a new and deeply disturbing element into the whole situation. My impression is that it was this factor which finally caused Jawaharlal to move away from the plebiscite idea, and instead start thinking in terms of accepting the *de facto* partition of Jammu and Kashmir. The joint communiqué of 20 August, therefore, marked the last chance of an agreed solution regarding the State between India and Pakistan.

Although Parliament, which was then in session, had listened to Jawaharlal's detailed report on the Kashmir events with interest and approval, and editorial comment in the country had been almost universally appreciative of the action we had taken against Sheikh Abdullah, in the valley itself there was a strong reaction. After the initial shock had worn off, a series of demonstrations erupted in various parts of the valley, especially in Srinagar and Anantnag, protesting against the dismissal and detention of the Sheikh. It took about a fortnight for Bakshi to restore normalcy. The police had to open fire on unruly mobs, but strict instructions had been given that the minimum necessary force should be used.

On 15 August I administered the oath of office to two more Cabinet Ministers, Pandit Shamlal Saraf and Syed Mir Qasim. The latter was a promising young lawyer from Anantnag, and I met him for the first time when he came to Karan Mahal to take the oath. I took a liking to him and we were to be good friends for many years thereafter. D. P. Dhar and Ghulam Rasool Renzu were sworn in as Deputy Ministers. Later, G. M. Sadiq, who was a major National Conference leader and President of the Constituent Assembly, also joined the Cabinet along with three more Deputy Ministers including the Head Lama of Ladakh, Kushak Bakula.

We had worked out a series of economic and administrative

measures to help counter the political reaction to Sheikh Abdullah's ouster. These included making all education in the State from the primary right up to the graduate level entirely free and abolishing the old practice of *Mujawazà*, the forced procurement of foodgrains from the peasants at fixed rates. Meanwhile, the National Conference under Bakshi held an impressive convention in Srinagar in September, and except for the hardliners, most of whom had been detained along with Sheikh Abdullah, the second-line leadership and the rank and file began to rally behind the new government. In Jammu and Ladakh, of course, there was relief and satisfaction at the change of government.

This was particularly so in Jammu. The bitterness and reservations that the people had nurtured about my accepting the office of *Sadar-i-Riyasat* disappeared overnight. Telegrams of congratulations poured in not only from Jammu but also from other parts of India, along with approving letters from unknown persons. Interestingly, these tended to interpret the dismissal in personal rather than political terms. The ones from Bengal almost invariably said that the death of Dr Shyama Prasad Mukherjee had been avenged, while a number of others congratulated me upon avenging the humiliation that Abdullah had heaped on my father by driving him into exile. My father himself was very guarded in his comments. He wrote in his next letter: 'All one can say and wish is that what has taken and is taking place may turn out to be for the good of the people and the State. That is also my prayer. You must be having a lot of strain in these exciting and busy days. Please take very good care of yourself.' Privately he sent me a message to tighten my security arrangements, perhaps fearing some attempted reprisal on the part of Abdullah's followers.

Bakshi was an entirely different sort of person from Abdullah. Despite minimal education—he never completed high school—he was an effective administrator. He owed this to his tremendous stamina, organizing capacity and genial public relations. Pragmatic in his approach to problems, and not unduly worried about the means that he used to achieve his objective, Bakshi, after an initial period of hesitation and even near-panic, rallied quickly and threw himself heart and soul into the new challenge. His genuine commitment to India was beyond doubt, and the role that he played in strengthening the national interest at that critical juncture must be recorded with gratitude.

For the first time since I became Head of State I began to feel that I was a real participant in the whole situation and not merely a figurehead. I instituted the practice of inviting Secretaries and Heads of Departments every week in batches of five over a cup of tea, so that I could acquaint myself with the administrative and developmental problems that the State was facing. Apart from keeping myself informed, those meetings gave me some insight into the actual modalities of administration which was useful many years later when I joined active politics and became a minister myself. I also started touring various parts of the State, particularly the Jammu region, meeting the general public and seeing at first hand the conditions under which most people lived. I thus began to emerge from the sheltered confines of the palace, and to open myself to a sharper awareness of the harsh realities of poverty and deprivation which were the lot of the masses. The affectionate and enthusiastic welcome I received during these tours revealed the regard in which my family was still held by the people, and formed the basis of my political career in the future.

Meanwhile, my inner quest continued. I kept occasional diaries during this period which I typed myself, and find that they are full of such references: 'What is this mystic experience about which all saints have written through the ages as being one of Absolute bliss and unadulterated joy? But how? How?' And again: 'I wonder when I shall also reach a stage of spiritual development in which I also might come into direct and intimate contact with the Lord of the Sacrifice, the Knower of the Field'; these terms taken, of course, from the teachings of Lord Krishna in the *Bhagavad Gita*.

These urgings were stimulated largely by Romain Rolland's *Life of Vivekananda*, Sri Aurobindo's *Synthesis of Yoga* and Paramahansa Yogananda's *Autobiography of a Yogi*. I was doing my daily puja regularly, but although it gave me strength it did not bring any of the mystic visions that so many seekers had experienced. However, my dreams became—and remain—

extremely rich in psychic material. Indeed, ever since my early twenties I have been having the most extraordinary dreams night after night, sometimes several in quick succession. It was only some years later after I had met Ṣri Krishnaprem that I realized that dreams, if properly interpreted, can provide valuable clues to one's psychic and psychological problems and to the manner in which the multiple crises of life can be surmounted.

The great appeal of Hinduism, of course, lies in its infinite diversity which caters to people with different psychological and spiritual backgrounds. Instead of trying to steamroll everyone into exactly the same formula or mode of worship, it allows the seeker a choice not only between a personal and a formless divinity, but between different aspects and images of the divine. Our family deity is Sri Rama, whose great temple in Jammu built by Maharaja Ranbir Singh is surely one of the finest in India. The main shrine is surrounded by a quadrangle which has one hundred and twenty-five thousand *shaligrams*, stones from the bed of sacred streams, then by a larger quadrangle with many temples containing images of almost all the major figures in the Hindu pantheon, and then again by the outer quadrangle which has Shiva temples enshrining the ashes of my ancestors. I have recently added to this a Nataraja temple, the first in north India, which has a lead crystal Shivalingam fabricated in Germany under which, in due course, some of my own and my wife's ashes will be placed.

But it was to Krishna that I was more drawn. Towards the end of 1953, after we had moved down to Jammu for the winter, I recorded in my diary: 'I have as yet not settled down to any one path or any one particular deity. Of course God is omnipresent, but nevertheless it is a help if one worships one particular aspect of Him. I am beginning to get greatly attached to the wonderful form of the Divine Lord Sri Krishna, and a deep and emotional love for him is beginning to grow in my being. ... I am now keenly awaiting some initiation into some form of yoga whereby I might really and truly set out upon the path of spiritual illumination.'

It was perhaps in search of such illumination that shortly thereafter I set out on an extended pilgrimage to east and south India. The quest for spiritual enlightenment has traditionally been linked in India with travel to various shrines situated in distant parts of the country. Until then, except for Punjab and Dehra Dun, Delhi and Bombay, I had seen hardly anything of India because I had been largely confined to my own State. Asha and I first visited my mother in the Kangra valley where she had rented a beautiful palace of the Nawab of Bhavalpur at a place called Alhilal between Palampur and Baijnath, which had been declared evacuee property. With her we went up to Mandi and Suket, small hill states in Himachal which have a great deal in common culturally with the Jammu region, and also visited Riwalsar, a place sacred to both Hindus and Buddhists, with a lake where small islands float around mysteriously. Then we proceeded to Calcutta, where I was astonished to see for the first time such vast numbers of people concentrated in a single city, more people than in the whole of Jammu and Kashmir.

We went to the great Kali temple at Dakshineshwar where Sri Ramakrishna, one of the most remarkable religious figures in history, had lived as the temple priest in the middle of the nineteenth century and had, through his astounding spiritual experiences, triggered off a virtual revolution within Hinduism at a time when a spiritual and intellectual crisis was sweeping across the land. I had read a good deal about him and his leading disciple Narendranath Datta, better known as Swami Vivekananda, who carried the teachings of his Master not only through the length and breadth of India but also to the Western world. I was fascinated by Vivekananda's personality, his glowing eloquence and leonine mien. The visit to Dakshineshwar was a moving experience, and on the return drive we went to the Belur Math, where Vivekananda founded the Ramakrishna Mission which carries on his work to this day. We also paid a visit to Kalighat, from which Calcutta derives its name, and saw there the goddess in her terrifying form. Terrifying, that is, for the evil-doers; for her devotees she remains the divine mother to whom so much of Hindu devotion is directed.

On the way back to Jammu we stopped at Delhi, and while we were there a tragedy occurred. Asha's uncle, General Bijaya Shumshere, was then the Nepalese Ambassador to India. We were at a morning cinema show when we got the news that he had been electrocuted in his bath by a defective immersion rod. We rushed to the embassy at Barakhamba Road where his body was lying. He was a talented and highly intelligent person, and despite the old family rivalry between the Ranas and the Shahs he had gained King Mahendra's confidence. Had he lived he would surely have made a mark in Nepali politics. Asha's father came from Kathmandu on hearing of the news of his younger brother's death, and their father Maharaja Mohun Shumshere, who had been living in exile in Bangalore since he left Nepal in 1951, also flew up to Delhi. It was a grim occasion when the whole Rana clan, along with the large circle of the Ambassador's Indian friends, assembled at the embassy. General Bijaya's son Pashupati was only about ten at the time. He proved to be the brightest among the younger generation of Ranas, and soon thereafter went to school in England in fulfilment of his father's wishes.

The visit to Calcutta only whetted my appetite for exploring the rest of India, and in 1954 I embarked on a long tour of the south including visits to many famous shrines. My mother, who had always yearned to do a *yatra* like this, also came with us along with her elder brother Thakur Nachint Chand and some servants. By the time we set off it was quite a large party, and we decided to go by train. It was truly a memorable visit for me, opening up aspects of our cultural heritage of which we in the north were hardly aware.

Here was the Hindu tradition in its pristine glory; the great temple *Gopurams* soaring into the sky covered with their incredible carvings often brightly painted; the richly caparisoned temple elephants, and the priests who received us at each of the great temples and conducted us through those massive monuments until we reached the inner sanctum where the main deity is installed; the haunting notes of the Karnataka music, so different from the north Indian mode and yet in some ways richer and more profound; the rhythmic, almost hypnotic chanting of Sanskrit prayers in unison, bringing out the glory of this greatest of languages in a deeply devotional setting: all combined to leave an indelible impression upon me. To this day I find myself attracted to the south, almost as if there was a spiritual link reaching far back into the recesses of time.

We visited the great Meenakshi temple in Madurai, dedicated to the fish-eyed goddess, and the Brihadishwara temple in Thanjavur with its huge Shivalingam, probably the most impressive temple in the whole of India. We saw the famous Shiva temple in Tiruvannamalai and visited the nearby ashram of Ramana Maharshi, one of the greatest saints of our times, who had passed away only four years earlier. We also saw the celebrated temple at Chidambaram where Shiva is worshipped in the form of the Divine Dancer—the Nataraja—one of the great artistic creations of mankind. We went to the temple of Srirangam in Tiruchirapalli, an entire town in itself, which we happened to reach on an especially auspicious day of the year—Vaikuntha Ekadashi—when the great northern gate is opened. We also went down to Rameshwaram, one of the four main pilgrim centres in the corners of India consecrated by Adi Shankaracharya, the remarkable eighth-century philosopher-saint who did so much for the regeneration of Hinduism and the spiritual unity of India. There Sri Rama, the hero of the *Ramayana* and for millions of Hindus God incarnate, had worshipped Shiva before setting out for Lanka, where his wife Sita was held captive by the demon king Ravana.

After the temples in Tamil Nadu, we proceeded further south to Kerala and went right down to Kanya Kumari, the tip of India where three great oceans meet. There we saw the small, exquisite temple dedicated to the Virgin Goddess, and a few miles away a huge statue of Hanuman at Suchindram. In Trichur we visited the famous Shiva temple built in the typical Kerala style of architecture, smaller and neater than the Tamil temples but in some ways even more appealing. As I visited these great shrines, redolent with myth and tradition yet vibrant with living devotion and uplifting ritual, it became clear to me that Hinduism was much richer than I had imagined living in the north ravaged by a thousand years of ruthless invasions, its temples destroyed, its statues ground into the dust. Here was the great reservoir of our culture and tradition, and I realized then that we of the north owe an undying debt of gratitude to south India for having preserved this great heritage intact down through the troubled centuries.

A special revelation was my introduction to Bharata Natyam, the unique dance-form based on Bharata's *Natya Shastra* which embodies the highest reaches of artistic achievement. Combining *bhava* (emotion), *raga* (melody) and *tala* (rhythm), the Bharata Natyam is the world's most impressive solo dance form. We were in Madras when we saw it for the first time; Vyjayantimala was performing in the Music Academy. I was stunned by the combination of swift movements and statuesque poses which she presented in a virtually continuous three-hour programme, at once deeply sensuous and spiritually inspiring. Ever since I have been a devotee of the dance, and have had the privilege of seeing many great artistes, including Balasaraswati and Yamini Krishnamurty whom I consider to have been outstanding in their respective generations.

During this visit we also went to Pondicherry, as I was becoming interested in the philosophy of that remarkable patriot-philosopher Sri Aurobindo and wanted to meet his spiritual collaborator Madame Alfassa. The Mother, as she was universally called, received us with great courtesy in her residence. I am afraid *my* mother was not terribly impressed by her, but I found a certain strange presence which bespoke unusual psychic power. I was to see her twice again before she passed away many years later, but it was Sri Aurobindo—whom I never saw in the flesh—who fascinated me. His fervent nationalism, mastery over the English language, soaring spiritual vision and comprehensive philosophy made a deep impression upon me.

The cumulative effect on my mind of this south Indian tour was considerable. It not only gave me a deeper awareness of Hinduism's many-splendoured dimensions, but also led me to realize that India was very much richer and more varied than I had imagined, and that despite its bewildering diversity it was Hindu culture that provided the unifying motif, the great backdrop, as it were, to our long and eventful history. While I did not meet a guru on this trip, it helped sharpen my awareness of the spiritual dimension which had, over the preceding years, been overshadowed by political events and activity. At the same time, I began to realize that I did not want to be indefinitely confined to one corner of India, even though that corner was as beautiful as Kashmir.

The Constituent Assembly had now resumed its meetings, but in vastly changed political circumstances. Whereas under

Sheikh Abdullah the State government was dragging its feet on the implementation of the Delhi Agreement except in those areas where it suited them, Bakshi now came out with a clear declaration that the agreement would be implemented in its entirety. During this period the Constituent Assembly was also functioning as the State Legislature, and as *Sadar-i-Riyasat* I addressed the Budget sessions. Instead of simply reading out a speech drafted by government officials I would ask for all the material and then put it into my own words. The speech was of course passed by the Cabinet before it was delivered.

Soon after assuming the office of *Sadar-i-Riyasat* I had secured the services of an able officer, Rajkumar Shivdev Singh, as my Secretary. For the post of Military Secretary I chose the genial Colonel Ghulam Qadir, better known as Harry Nedou, who was Begum Abdullah's brother. Despite Sheikh Abdullah's dismissal he continued in his post as he was entirely non-political. An excellent shikari, he used to organize shooting and fishing trips for us and was greatly disappointed when I gave up both these activities. My aides continued to be drawn from the former Jammu and Kashmir State Forces, which at my persistent suggestion were not disbanded as in other States but integrated *en masse* into the Indian army as the JAK Regiment. This continues to the present day to play a notable role in the defence of the nation and, along with the Dogra Regiment, provides welcome job opportunities mainly to the Dogra community, which has a long and proud martial tradition.

Some other steps were taken to strengthen the links between the State and the rest of India. On Baisakhi day, 13 April 1954, the customs duties on goods entering the State were abolished, which immediately had the beneficial effect of lowering prices of essential commodities and also created an important psychological impact on the people as it removed yet another barrier. During this period Dr Rajendra Prasad, at my invitation, paid an official visit to Jammu. He was given an affectionate reception by the people irrespective of their political differences. The image of gracious self-effacement, Rajen Babu made a deep impression upon those with whom he came into contact by his genuine goodness and humility. I escorted him to the important temples in Jammu where he worshipped with great devotion. He also inaugurated the Jammu and

Kashmir State Forces War Memorial Hall in the Jammu Cantonment at Satwari; the function was attended by thousands of ex-servicemen including grizzled veterans who proudly displayed their medals on fading shirts. The President and his party stayed as my personal guests in Hari Niwas; and for many years thereafter whenever we met, Rajen Babu would refer with pleasure to the days he spent there enjoying the spectacular view of the mountains across the river Tawi.

The process of strengthening the Centre–State relationship now began gathering momentum. On 14 May 1954 the President promulgated an order extending various articles of the Indian Constitution to the State as requested by the State government. By then we had moved up to Srinagar for the summer, the curious biannual migration to and from the valley which has become so much a part of life in the State that it is looked upon almost as a natural phenomenon. A great deal of time and money is involved in these moves, but they reflect the bifocal nature of Jammu and Kashmir State. The third and by far the largest region, though very sparsely populated, is Ladakh, but this remained remote and largely neglected although Kushak Bakula's entry into the Council of Ministers did give Ladakh political representation for the first time.

In 1954 the eminent hotelier M. S. Oberoi (known universally as 'Rai Bahadur') approached me through D. P. Dhar with a suggestion that I should lease to him Ghulab Bhavan, our main palace in Srinagar, which was lying virtually unused. I thought that this was an excellent idea, because even though the lease money was very modest, the property would be well maintained and would provide Kashmir with its first five-star hotel. I was the first of the former princes to take this step. Jaipur and Udaipur followed a few years later, and now, of course, it is well recognized that these massive palaces dotted throughout the country can play an important role in strengthening the tourism infrastructure.

As a result of a directive from Jawaharlal, Srinagar became the venue of numerous conferences organized by ministries and departments of the Government of India as well as private organizations. We always had a well organized *Tawaza* (hospitality) department, and under Bakshi the entertainment of official visitors was raised to the level of a fine art. An aggressively hospitable man, with the disconcerting habit of stuffing *goshtabas* (Kashmiri meat balls) into the mouths of his unsuspecting guests before ascertaining whether or not they were vegetarian, Bakshi was always the life and soul of the party. While this was all right to begin with, the lavish hospitality in later years tended to become a subtle sort of favouritism to persons from whom special favours were expected. Indeed Bakshi's '*goshtaba* hospitality' has passed into the phraseology of Kashmir.

I decided to restart my studies, and Asha followed suit. She began preparing for the Bhushan examination in Hindi, which is equivalent to an intermediate course. She also started a branch of the All-India Women's Conference and became President of the State Social Welfare Board. On my part I decided to study for the M.A. examination in Political Science, this time in Delhi University. As a special courtesy, because I was chancellor of a sister university, Delhi University allowed me to study privately and come to Delhi only for the examinations. The M.A. course was divided into two examinations—Previous and Final—each with four papers covering Indian and Western Political Thought, Indian and Comparative Constitutions, International Relations and International Law. For guidance in these subjects I turned again to Professor P. N. Chaku, Dr J. N. Bhan, who had recently returned from the London School of Economics, and a younger teacher, Professor Satya Bhushan, who was from Jammu.

The constant stream of visitors from all over India and the world continued, as did our regular lunch and tea parties. Among those who came up that year was Gandhiji's close associate Mirabehn (Margaret Slade). A large, rather formidable lady dressed in an orange sari, she had started a cattle farm near Kangan in the Sindh valley, where she installed a herd of tiny Dexter cattle imported from England. These gatherings, though essentially social functions, gave me the opportunity to converse with a wide variety of people drawn from various walks of life, and I learnt a great deal from them. Asha had by now become quite fluent in English and began participating increasingly in the conversation.

Meanwhile, developments within political parties at the national level were taking place. Rafi Ahmed Kidwai, who had been the main adviser to Jawaharlal on several political matters including Kashmir, died suddenly of a heart attack. Asoka Mehta and some of his colleagues from the Praja Socialist Party toured the State and attempted to extend national politics to Jammu and Kashmir. A branch of the party was set up in the State with Om Prakash Saraf as President and Khwaja Umar Butt as Secretary. Simultaneously, the Praja Parishad announced its affiliation to the Bharatiya Jana Sangh. Thus, while the Congress and the National Conference retained close fraternal ties, the process of political-party integration between the State and the rest of India gained momentum. Although this was not welcomed by some Congressmen, who tended to judge every issue in narrow party terms, many of us felt it to be a healthy development as it forged further links at the political level between the people of the State and the rest of India.

In my view all links that strengthened this relationship were to be encouraged. This led me to write a long letter to Jawaharlal on 25 July 1954 making a strong plea for the construction of a road linking Leh with the Indian mainland via Manali and the Rohtang Pass. I marshalled strategic, economic and political arguments in favour of the proposal and *inter alia*, said:

> For over a century Ladakh has been part of the Jammu and Kashmir State, and for about seven years part of India. I feel that its people have loyalty and affection for India, and left to themselves they will remain loyal. But in view of recent developments on the 'Roof of the world' it is most important that these ties with India must be strengthened so that they become virtually unbreakable. Uptil now the authorities in Tibet have not shown too much interest in Ladakh, but there is no guarantee that this attitude will continue indefinitely, and after consolidation in Tibet it will be too optimistic to believe that Ladakh will escape an undue share of attention. Our information, meagre as it is, shows that the Chinese are fast completing the road linking Lhasa with Peking thousands of miles away, and are also engaged in constructing a network of roads within Tibet.

Within a week came his brief but positive reply, dated 28 July 1954:

> My dear Tiger,
>
> Thank you for your letter of the 20th July.
>
> You have advanced powerful arguments in favour of a road from Manali to Leh. The proposal attracts me greatly and I wish I could

take it up. The difficulties, however, are formidable. Nevertheless, I am having this matter looked into again and, so far as I am concerned, I should like to have this road if at all possible.

Yours ever
Jawaharlal Nehru

The survey of this road, so vital for our national security, was begun in right earnest, and in due course it was taken in hand and completed. However, this came too late to prevent the tragedy of 1962 when China invaded India, which still lay eight years in the future. The first commercial air service by Indian Airlines from Srinagar to Leh was inaugurated soon thereafter on a fortnightly basis, but this provided at best a tenuous, symbolic link with the most vulnerable region of Jammu and Kashmir State.

On the railway front too we had been completely cut off since partition, which severed the Jammu–Sialkot rail link. At that time the nearest railhead was in Pathankot, but a decision was taken towards the end of 1954 to extend the line to Lakhanpur, which was the border between Jammu and Punjab. Work was started on this, but it took many long years for the railhead to finally reach Jammu. Meanwhile, an ambitious project of drilling a tunnel through the Banihal Pass which would enable the Jammu–Srinagar highway to be kept open around the year was also started with German technical collaboration. Taken together, events in 1954 saw the stepping up of efforts to forge closer bonds between the State and the rest of India. I sincerely felt that this was as much in the national interest as in the interests of the people of the State themselves, and I did whatever I could to encourage these activities.

There were, of course, some countervailing developments also. Sheikh Abdullah was still in detention, but his closest colleague, Mirza Afzal Beg, and some other detenus had been released and had restarted a measure of political activity in the valley, mainly in Srinagar city and his home district of Anantnag. Indeed, he founded an opposition group of about eleven people in the Assembly, but the bulk of the National Conference members remained loyal to Bakshi and no real threat to the government was posed.

An agitation was started on behalf of the Mirpur Refugee Committee asserting that the government had not fulfilled its promises to provide the post-1947 refugees with adequate compensation and rehabilitation benefits. This brought to light a curious contradiction in the Government of India's policy towards Jammu and Kashmir refugees which continues down to the present day. While refugees from the erstwhile west Punjab received full compensation, those from the Pakistan-occupied areas of Jammu and Kashmir were denied this on the ground that as their territory was legally part of Jammu and Kashmir State, to treat them as refugees would weaken our international position. Thus, for no fault of theirs, our refugees were reduced to a second-class status even within the refugee community. Added to this was the fact that about ten thousand families of Hindus (mostly Harijans) came into the State from across the Sialkot border to evade the massacre of minorities there, and they were in an even worse position because they were *not* considered to be State subjects at all. So, in effect, the refugees in our State got a doubly raw deal, and although I took up the matter on innumerable occasions with the Government of India we were not able to cut through the legal-cum-bureaucratic rigidity and thus failed to do full justice to these unfortunate people.

Meanwhile, my discovery of India continued. After addressing the Convocation of Punjab University at Hoshiarpur I went on to visit the gigantic hydro-electric project at Bhakra-Nangal. It was astonishing to see the dimensions of the dam being constructed there, and I felt a sense of pride that so soon after Independence we were able to undertake such an impressive project. I was also beginning to get interested in the world outside India, and when I read in the papers that Jawaharlal was visiting the Soviet Union I wrote and asked him whether I could come along. Characteristically he replied immediately on 21 March 1955:

My dear Tiger,

Your letter of the 19th March.

I quite understand your desire to go to the U.S.S.R. but I do not think it will be feasible or desirable for you to go with me. I am going in response to a special invitation. I take a small staff with me to do my work. Otherwise, it will not be proper for me to add any important person to my party. Also, your going would have a certain political significance which might not be good for Kashmir at present.

I hope, however, that you will have a chance of going there later. There is no particular difficulty about it.

Yours ever
Jawaharlal Nehru

While I did not get to visit the Soviet Union that year, in a way the Soviet Union came to visit us instead. In November 1955 the celebrated visit of Bulganin and Khrushchev to India took place. We had followed with great interest the bold liberalization moves initiated by Khrushchev after the death of Stalin. The oppressive nature of Stalin's regime had been widely commented upon, but many tended to write this off as part of the US-sponsored anti-communist propaganda campaign. However, when Khrushchev succeeded Stalin after a brief but bloody confrontation with the Security boss Beria, and himself denounced the terrible excesses of the previous regime including the slave camps in Siberia, it became clear to all but the most dogmatic Stalinists that the people of Russia had paid a terrible price for their material progress under that regime.

During Stalin's time the Soviet attitude towards India was one of suspicion and overt hostility. Even Jawaharlal was, incredibly enough, described in an official party publication as 'the running dog of Western imperialism'. However, after Khrushchev's assumption of power a distinct change came over the Soviet policy in this regard, and his visit (along with the titular Head of State, Bulganin) marked the beginning of a new phase in Indo-Soviet relations. Asha and I happened to be in Delhi just before the visit and at the suggestion of my aide Captain Govardhan Singh we stayed on for a few days. When the Prime Minister learnt that we were still there he invited us to dinner at his house where we were introduced to Khrushchev, who insisted that Asha and I looked like brother and sister! In the course of our conversation I suggested that Khrushchev should pay a visit to Kashmir, and he turned to Jawaharlal with a smile.

Immediately after I returned to Jammu I received a letter from the Prime Minister dated 21 November to say: 'You spoke to me about Bulganin and Khrushchev going to Kashmir. Later this evening they told me that they had decided to go there and left it to me to fix a suitable day. This is not an easy matter, but of course we shall have to find some way.' After adding some more details he said: 'They want to spend one clear day in Srinagar. That will mean really two nights and a good part of two days. I shall keep you informed.' Two days later he wrote again giving details of the visit and adding: 'I think you should go to Srinagar for this occasion. In fact, it was largely due to your insistence that they agreed to go there.' Several other letters followed, which showed the keen personal interest Jawaharlal was taking in the visit. He was worried that the rich Kashmiri food might upset them. At one point he wrote:

> Information reaches me that both Bulganin and Khrushchev are on the point of collapsing because of the strain of the heavy programme. They have asked me to cut out many items from their future programmes, and I am doing that. This means that, in Srinagar, their programme must be very light indeed.

Though both the Soviet leaders were characteristically square and tough, the long tour, including a trip to Burma, had evidently begun taking its toll.

Finally, all was settled. Asha and I went to Srinagar and on 9 December both the leaders arrived. We received them at the airport along with Indira Gandhi, who had flown up a day earlier. They came in a special plane and, according to instructions from Delhi, were given co-equal protocol treatment for the guard of honour. Evidently, even though Bulganin was Head of State, effective power was being wielded by Khrushchev. After the elaborate reception at the airport we went towards my convertible Chevrolet for the procession into town. Both the leaders had to sit behind, and as the *Sadar-i-Riyasat* I had to do the same. The result was one of the tightest journeys I have ever had. Khrushchev had a hearty sense of humour and roared with laughter as he told me of a similar ride with Maharaja Jaya Chamaraja Wadiyar (who was about three times my size) during their visit to Mysore.

The whole Soviet party was put up at Gulab Bhavan, now the Oberoi Palace Hotel, and Indira Gandhi and the entire Delhi team also stayed there. The interpreter from our side was a brilliant linguist, Amina Ahmed, daughter of the Delhi Mayor Nuruddin Ahmed. Vishnu Ahuja, then a junior officer in the Foreign Ministry and also a Russian expert, was on the team. Whether they conducted their courtship in Russian I have never been able to discover, but they were married soon thereafter and Ahuja finally retired as our Ambassador in Moscow.

The next day, as suggested by Jawaharlal, we gave the visiting dignitaries lunch at Karan Mahal. They were enthralled with the double view from there, and left lovely gifts of Russian ceramic figures—the King of the Sea and the Queen of the Sea—which still adorn our foyer. When I remarked that I thought the communists didn't believe in kings and queens, Khrushchev laughed uproariously and said that they were all right as long as they were ceramics!

In the evening there was a civic reception at the Rajgadh Palace on the Jhelum, where, under the chinars in the crisp winter air, Khrushchev delivered the famous speech when, for the first time, the Soviet Union came out openly in support of Jammu and Kashmir being an integral part of India. Referring to the proximity of the Soviet Union he said, 'If you are ever in difficulty all you have to do is to climb on top of these mountains and shout, and we will come to your aid.' The same sentiments were echoed by the Soviet leaders at the final function at the airport on the 11th in response to my farewell speech. Thus ended a historic visit, and my first exposure to international affairs.

Chapter 14

Early in 1956, on the advice of the Government of India, Bakshi Ghulam Mohammad decided that Sheikh Abdullah would be released. This naturally caused a good deal of trepidation among those of us who felt that, despite the useful measures adopted by the Bakshi government since August 1953 and the welcome steps taken towards strengthening the ties between Jammu and Kashmir and the rest of India, the political situation was still not stable enough to warrant taking the risk. The law and order situation could deteriorate rapidly, and as the Constituent Assembly was still functioning it was even possible that Sheikh Abdullah could win back a sufficient number of members to topple the Bakshi regime. I decided to write in detail to Jawaharlal frankly putting forward these apprehensions. This I did on 11 January 1956, stating, *inter alia*: 'It has of course to be considered how long he can be kept in detention. I support the widely held view that the proper time to release him, keeping in view the security of the State which is our primary responsibility, would be only after it is found possible to declare that the Kashmir dispute is finally closed, or at least until the State Constitution has been completely enacted and the Constituent Assembly dissolved.'

I sent the letter to Delhi through my ADC, and Jawaharlal sent a long reply on that very day, 11 January. He started by saying: 'During the last year or more, the question of Sheikh Abdullah's release has often been discussed. Obviously it is a difficult question and one has to balance various factors. So far as I am concerned, my whole mind rebels against the long detention of any person without trial. I have objected to this so often in the past that naturally I do not like it. But I realize that sometimes circumstances compel one to take action which is normally undesirable. In the balance, therefore, I left it to the judgment of the Jammu and Kashmir Government to decide what they should do in the matter.'

After several paragraphs he stated: 'I realize fully the risks involved. But one does not solve a problem or really avoid risks by running away from them. Therefore, after giving a great deal of thought to this matter, I have felt that Bakshi Ghulam Mohammad should take the action he has decided upon. This is as favourable an opportunity as is likely to occur. Of course we should be fully prepared. Having got over this difficulty, the future of the Kashmir problem will be very much simpler.'

This was the last typed paragraph of his letter. Thereafter, significantly, he added two paragraphs in his own handwriting, which were as follows:

> I would add that the visit of the Soviet leaders to Kashmir and its effect on the internal and external situation, which has been all to the good and has influenced many people in the right direction, is a factor to be considered. The conditions in Pakistan and in 'Azad' Kashmir are bad. Sheikh Abdullah may behave very foolishly but the effect of his behaviour is likely to be less now and can be dealt with more easily.
>
> All these are speculations and appraisals of a difficult situation which has to be faced now or later. The question is what the most suitable time will be. It is bad to live in apprehension all the time. It is better to take the ghost out of our minds and deal with it.

Another matter about which I wrote to Jawaharlal at this stage, in keeping with my policy of corresponding with him on matters that I considered to be of national importance *vis-à-vis* the State, was the question of the States' Reorganization Commission's Report. While Jammu and Kashmir was kept outside the terms of reference of the Commission, I brought to Jawaharlal's notice three points that I felt deserved special consideration. The first was that Jammu and Kashmir should at least be included in the Northern Zone which had been envisaged as part of the zonal set-up recommended by the Commission. Secondly, the SRC's recommendations regarding linguistic minorities should be applied to Jammu and Kashmir. I wrote: 'As you know, there are two important linguistic minorities in the State, the Dogras and the Ladakhis, and I am sure that no one will want them to be deprived of the safeguards that are to be provided for other linguistic minorities in the rest of India.' Thirdly, I urged that the President of India should be empowered to nominate the Head of State of Jammu and Kashmir, whatever that office may be called. I added that 'it does not matter who is the Head of State here so long as we can be sure that it will always be someone who will keep the larger national interest paramount and enjoy the complete confidence of the Centre.'

Having worked over the last year for several hours a day at the M.A. course, I went down to Delhi early in April, a month before the M.A. (Previous) examination and stayed at Maiden's Hotel, which is located conveniently near the University. The examination was conducted in Delhi University, where I was allotted a seat in the hall along with about fifty other students. Because of my stiff hip I had difficulty in sitting on the normal low chairs, and the only concession made to me was to provide me with a higher chair and table. Each paper was for three hours, and because I had done my preparation thoroughly, including a considerable amount of supplementary reading outside the prescribed course, I was able to write my answers with confidence; the combined results were to be announced only after the final examination the following year.

While in Delhi I also began looking into the construction of our new house in Chanakyapuri. I had acquired a one-acre plot on the suggestion of the then Minister for Works and Housing, Sardar Swaran Singh, and had asked a brilliant architect, Durga Shankar Bajpai, son of the distinguished civil servant Sir Girija Shankar, to design the house. At that time Chanakyapuri (the Diplomatic Enclave) near Ashoka Hotel was virtually a wasteland, and some friends thought that it was most unwise for us to have decided to build a house in the wilderness. Today, of course, it is one of the most remarkable residential enclaves of the world, and with each country having designed its own mission it has become a living museum of contemporary architecture.

Durga designed a lovely house for us, which was to prove a boon in the years to come. Most unfortunately, he came down with meningitis while the construction was still in progress, which left him an invalid for the remaining ten years of his life. It was all the more tragic as Durga was one of the most talented architects of our post-Independence generation. The house, which we called 'Mansarovar' because it was built round a pool, was subsequently completed by Durga Bajpai's friend and professional associate Piloo Mody. Both Durga and Piloo were a few years senior to me at the Doon School.

Meanwhile the trip abroad which we had planned ever since we were married in 1950 had still not matured. We had hoped to go in 1953, but in view of the political crisis that year had to call it off. In 1956 we decided to go for about two months. I discussed this with Jawaharlal and Bakshi. After some consultation with constitutional advisers it was decided that there was no need for anyone to be appointed to perform my duties while I was out of the State, and that if there were any important papers needing my attention they would be sent to me wherever I was. I decided to take Dr J. N. Bhan with us, because he had lived in London for several years and would be helpful on the trip. Some eyebrows were raised among the bureaucracy in Delhi because of his leftist leanings, but no serious objection was taken.

My father had been living outside the State since 1950. For most of the year he would live in the top floor of a block of flats called 'Hari Bhawan' that he had constructed at 64 Pedder Road in Bombay, but in the monsoons he would shift to Poona for a couple of months for the racing, and in the winter go down to Bangalore for the same purpose. Always a keen racer, his life now revolved almost exclusively around this sport and he continued to own a string of excellent horses. We spent a few days with him in Bombay before starting our discovery of Europe. Always neat and meticulous, he made out detailed lists of the items Asha and I should take with us on our trip, weighed our suitcases after they were packed to ensure that we were not exceeding the weight allowed by the airline. Both Asha and I were excited about this visit to which we had been looking forward for years.

We have now travelled hundreds of thousands of miles, virtually to the ends of the earth. But that first holiday had a very special quality—the excitement about the great works of Western culture mingling with the romance of a second honeymoon. Our first stop was Cairo, where we gaped at the ageless Sphinx and the massive Pyramids behind it, relics of a great civilization that pre-dated even the Indus Valley. The museum in Cairo was fascinating, except that the 'mummies' were a bit of a let down. In fact an ancient woman was sitting on a bench nearby, and Asha, mistaking *her* for a mummy, got a nasty shock when she slowly rose and wobbled away. The ancient Egyptian civilization was indeed one of the high-water marks of human achievement, but its obsession with death and *physical* immortality, as reflected in the intricate tombs and in the art of embalming, is rather intriguing. While in Cairo we also visited that masterpiece of Islamic architecture, the Mohammad Ali mosque, made of alabaster. It is a magnificent structure characteristically eliminating the depiction of human figures in keeping with the tenets of Islam. The whole artistic effort is channellized into elaborate carving of skill and beauty in which verses from the Holy Quran are inscribed.

After Cairo our European odyssey began. Our first stop was Italy, where we were absolutely stunned to see the great churches in Rome, the heart of Christendom. St Peter's Cathedral is surely the most beautiful building in the world from the inside as the Taj Mahal is, in my view, the most perfect from the outside. Entering the great church one gets some idea of the power and glory of Christianity, and the vitality that enabled this religion to play such a crucial role in world history. The 'Pietà', a marvellous life-size sculpture by Michaelangelo showing the body of Jesus just brought down from the cross in the arms of his mother Mary, is perhaps the most moving piece of art anywhere in the world, and stands to the right as one enters the cathedral. Then there is the Sistine Chapel, sculpted out of rare marble by the same master-artist, the exquisite paintings on the ceiling of the great dome that Michaelangelo executed lying on his back for years, and a mosaic showing the baptism of Jesus so beautifully that the water looks absolutely real.

We also saw several other important churches in Rome including the one dedicated to Saint Paul, whose genius really founded the Catholic Church on the firm rock of Saint Peter. In a smaller church was another superb statue by Michaelangelo showing Moses just descended from Mount Sinai with the tablets in his arms, so realistic that one could swear he was breathing. We had a marvellous guide called Hazan who told us all the myths and stories connected with the sculptures. It is said that after creating Moses, Michaelangelo looked at him for a long while and then shouted, 'Moses, speak to me.' When the statue did not respond he threw his chisel at it, and a small notch marks the impact. Such is the divine madness of genius.

A third great sculpture of Michaelangelo is the gigantic figure of David, over sixteen feet high, that stands in the main square of Florence. The perfect proportions of the nude body make this a most extraordinary piece of art. Asha was enchanted with the Sunday bazaar held in the nearby hills where villagers sell their handicrafts and wicker baskets. Asha herself looked stunning in her colourful saris, and we could hardly move in an Italian town without eliciting good-natured whistles. The Italians are great admirers of female beauty and make no secret of it. What amazed me was the way they sat late into the night in open air cafés, drinking coffee or sipping wine as if they had no care in the world. From Florence we went on to Venice, that unique oceanic city with canals instead of roads, with its St Mark's Square which is one of the most picturesque anywhere in the world. Stepping out of the train we found ourselves on the main canal, from where we took a gondola to our hotel. Though much exploited for tourist traffic, Venice remains one of the two most romantic towns in the world, the other being the golden city of Jaisalmer in the Rajasthan desert. Thus the artistic genius of man has transformed both ocean and desert into gems of architectural beauty, poised for a brief moment of time on the crest of eternity.

From Italy we went on to Frankfurt and Dusseldorf in the Federal Republic of Germany, where we shopped on the Kaiserstrasse and took a fascinating trip in a boat down the Rhine to Cologne, where we saw another great cathedral, miraculously undamaged despite the saturation bombing of Germany during those last dreadful months of the Führer's satanic regime. Then we went to Switzerland, which remains to this day my favourite country outside India. On that first trip we went to Geneva and then to Berne, where we saw the polar bears in their icy cages. On the ropeway up to Jungfrau it was cloudy all the way until, at the very top, the clouds broke and the great peak—the 'Young Lady' of the Alps—burst upon us with all the splendour of a revelation.

On from there to Paris, where Victor Rosenthal had laid on a lavish reception. Asha and I did the sights of Paris, and then came a high point of this visit, the trip to my birthplace in Cannes, where I had arranged to stay in the Hotel Martinez in the very suite in which I was born on 9 March 1931. M. Martinez, the proprietor, was still alive and active at the time and greeted me warmly. 'You look exactly like your father,' he kept exclaiming. We had some portraits taken by the well-known photographer J. Stara in Cannes, went for a cruise on the blue Mediterranean, and visited St Tropez, Monaco and Nice close by.

Finally we proceeded to England accompanied by Beryl Stileman, who had come over to France to be with us. As always, she was full of charm and vigour, and it was great fun having her with us. She and her husband Freddie had been dear friends of my parents way back in the racing days in Bombay before Independence, and they had returned to England soon thereafter. Freddie died a few years later; Beryl, now ninety, remains a close friend. She had organized an interesting programme for us including visits to the House of Commons (where Edward Heath, then Parliamentary Secretary, gave us lunch); Oxford (where we ambled around some of the colleges and met several dons) and Wimbledon (where in the men's finals we saw the Australian Rod Laver beat his compatriot Roy Emerson). In London we saw several plays including Agatha Christie's *Mousetrap*, which went on to become the longest running play in the world, and a musical called *Kismet*. We also attended a garden party at Buckingham Palace where the 'Blue Danube' was being played by an impressive band. By association this glorious waltz became a sort of theme for the trip; it was playing almost everywhere we went. We were entertained to dinner by the gracious and highly intelligent sister of Jawaharlal Nehru, Vijayalakshmi Pandit, who was our High Commissioner.

All in all, this visit was as enjoyable as it was educative. Having moved out of my shell in Jammu and Kashmir and begun a discovery of India, this gave me an extra dimension, an awareness of the world *outside* India that helped to put our own development in a clearer perspective. Having started travelling when I was only six weeks old, I can claim to have what in India is called 'wheels on the feet', the propensity to travel to the ends of the earth. Apart from the polar zones, there is hardly any major continent that I have not visited in these years, and seeing new lands invariably rejuvenates me. With the world now a much smaller planet than it ever was, and shrinking rapidly with every advance in travel and communications technology, travel brings to mind sharply the concept of our ancient seers who looked upon mankind as a family—*vasudhaiva kutumbakam*. Indeed, if human civilization as we know it does manage to survive into the twenty-first century, there is even the off chance of my being able to visit the moon.

By the time we returned from abroad, the drafting of the State Constitution by the Constituent Assembly had gathered momentum. I had secured a copy of the first draft, and after studying it wrote to Jawaharlal towards the end of July enclosing my detailed comments on various provisions. Among other points I suggested that instead of having a separate Election Commission, the National Delimitation Committee should be entrusted with this work, especially in so far as it concerned parliamentary elections. Till then the six MPs to the Lok Sabha from Jammu and Kashmir had been nominated by the State Assembly, but I saw 'no good reason why our people should be indefinitely deprived of the privilege of directly electing their representatives to Parliament'. I also supported the Punjab regional formula as a pattern for Jammu/Kashmir relations *inter se*, and urged special provisions for the huge but sparsely populated frontier region of Ladakh.

The draft Constitution was in fact introduced in the Constituent Assembly on 10 October 1956, and it met for clause-by-clause discussion on the 22nd. In order to give them an opportunity to participate in the debate, all the detained members were released with the exception of Sheikh Abdullah. They attended the session led by Mirza Afzal Beg, who launched a strong attack upon the government and questioned the very legality of our accession to India. He also made intemperate speeches outside the Assembly, virtually exhorting his followers to fight a last-ditch battle against the passage of the Constitution, even to the extent of physical violence against prominent government personalities. As a result of these intemperate attacks Mirza Beg and his colleagues were again detained. From jail Abdullah wrote a long, irate letter to G. M. Sadiq, the President of the Constituent Assembly, fiercely attacking all that had occurred since his detention in 1953. At the other end of the political spectrum the Praja Parishad launched a campaign against the Constituent Assembly on the grounds that a separate Constitution for the State was not necessary at all and that the Indian Constitution should apply in toto.

Despite all this thunder from both sides, the Constituent Assembly proceeded with its work and on 17 November the members put their signature to the new Constitution, thus marking a new phase in the chequered history of Jammu and Kashmir. The Constitution was passed unanimously by the 68 members present under the Presidentship of G. M. Sadiq, a veteran and highly respected National Conference leader. Seven members were under detention, including Sheikh Abdullah and Mirza Beg. Some provisions came into effect at once, but the bulk were to take effect from the next Republic Day, 26 January 1957. It was also planned that general elections in the State would be held that year along with the rest of the country. For me these developments were a source of considerable satisfaction, as they marked an important strengthening of the constitutional, administrative and political ties between the State and the rest of India.

We visited Delhi towards the end of November and were invited to participate in functions connected with the historic first visit of the Chinese Premier Chou En-lai along with the Dalai Lama and the Panchen Lama. At a banquet in Ashoka Hotel Jawaharlal took me up to Chou En-lai and introduced me to him with some brief remarks about my Kashmir connection. I found the Chinese Premier suave and smooth, his smiling countenance revealing little of his feelings. I was more interested in meeting the two famous Tibetan Lamas, only a few years younger than I, of whom we had heard so much. The Dalai Lama was an unusually impressive person who seemed even at that age to have a spiritual dimension. Little did I then realize that he was soon to become the tragic symbol of conflict between the world's two most populous nations, and between the two eminent Asian statesmen who exchanged toasts with such cordiality in Ashoka Hotel that evening.

Soon after I got back to Jammu an event of importance took place in the inauguration of the three-kilometre tunnel through the Banihal mountain, a marvel of modern technology, which had been completed in record time with the collaboration of a firm of German engineers. Boring had begun from both the Kashmir and the Jammu ends simultaneously, and it is said that the German Chief Engineer slept with a loaded revolver under his pillow on the last day when the two sides were to meet in the middle of the mountain, so that if there had been any wrong calculation and the tunnels missed each other he would shoot himself. Happily this gory contingency did not arise, the tunnels were perfectly aligned and the whole project represented a major communications breakthrough. Previously traffic had to go almost to the top of the 9000-foot Banihal Pass, the approach to which was blocked for three to four months a year depending on the extent of snowfall. The new tunnel was constructed at an elevation of 7200 feet, which meant that with the help of snow ploughs the Jammu–Kashmir highway could be kept open throughout the year for civilian as well as military traffic.

We decided to call it the Jawahar Tunnel because Jawaharlal had approved and supported the project, and we had wanted him to inaugurate it. He declined due to heavy commitments, so I suggested that we invite Dr S. Radhakrishnan, the brilliant scholar who was then the Vice-President, to perform the inauguration. This took place on 22 December, and Dr Radhakrishnan, Bakshi and I packed into my trusty convertible and drove up to Ramban, half way between Jammu and Srinagar. From there we proceeded in a colourful procession to the site of the tunnel, where a tape was cut and speeches made. Then we drove through it, a curious experience because of the sensory distortion caused by the prolonged darkness and lack of fresh air. After what seemed a very long time, but was in fact only a few minutes, we burst into bright sunlight on the Kashmir side, made even more brilliant by its reflection off the deep snow piled on either side of the road. There was great rejoicing at both ends of the tunnel, and thousands of villagers had turned up to witness the historic event.

It was always a joy to hear Dr Radhakrishnan speak, whether at an academic or a public function. Without a scrap of paper, he would always make a most impressive speech, his crisp enunciation punctuated with appropriate Sanskrit quotations. He had a clear and logical mind, and would present his ideas in brief but effective language. It was such a welcome change from the usual run of public figures in India, who generally had little to say and took a painfully long time saying it. Dr Radhakrishnan in fact took a liking to me, especially because of my interest in the Upanishads and the *Bhagavad Gita*, over which he had such mastery. Whenever I went to Delhi I would call on him and we would have long chats. Despite his great learning he had a certain engaging pride. On the return journey from Banihal he turned to me with a wry smile and said, 'You invite me to open the tunnel and then name it after Jawaharlal Nehru.'

Early in the new year the whole Kashmir issue came up again in the Security Council. Pakistan had evidently panicked at the constitutional developments in the State. It observed a 'Kashmir Day', stepped up its hostile media campaign and on 16 January 1957 the Security Council met to discuss a letter written by the Pakistan Foreign Minister Feroz Khan Noon. Noon himself presented the Pakistani case, while the Indian case was put forth by Krishna Menon in one of his famous marathon speeches that took three meetings to complete. Never before had our case been presented in such a detailed and elaborate manner, replete with quotations from documents and press reports. The main point that Menon put across was that Kashmir was not a territorial dispute but a clear case of aggression by Pakistan, and that the Security Council must take 'action consistent with the crime of invasion'. Regarding our Constituent Assembly he met the Pakistan tirade by saying that its actions would be declaratory, not creative, because it was a 'sub-sovereign' body.

Despite this, the Security Council adopted a resolution reaffirming the earlier UN and UNCIP resolutions, and also reaffirmed the resolution of 30 March 1951 that the convening of a Constituent Assembly and any action taken by it would not constitute disposition of the State as envisaged by the UN. Significantly, the Soviet Union alone opposed these views and stated that the adoption of the Constitution by the State Constituent Assembly showed that the people of the State considered it 'to be an inalienable part of the Republic of India'.

Despite the passage of the Security Council resolution, the Constituent Assembly met at night on 25 January for its concluding session. At midnight the new Constitution came into effect and the Constituent Assembly was dissolved after having completed the work for which I had convened it five years earlier in 1952. It continued to sit as a legislative body for a few days to pass a vote-on-account budget and several bills including the Representation of the People's Act which was necessary for the holding of the elections. It also promulgated a new set of Business Rules to bring the functioning of the government machinery on to a more rational and systematic basis.

These positive developments, however, were accompanied by a marked increase in tension in the valley. The Security Council Resolution did put new life into the pro-Pakistan forces, and the Pakistani delegate produced what purported to be a letter from Sheikh Abdullah to members of the Security Council, which caused quite a sensation. Wild rumours began circulating of a possible invasion of the State by Pakistan on 26 January to coincide with the adoption of the Constitution. In the event no such situation materialized until several years later.

While all this excitement was in progress, there were important developments on the domestic front also. When we were married in 1950 Asha was only thirteen and although both my mother and father were desperately anxious for a grandson—a good old Hindu tradition that shows no sign of waning—it would have been really absurd for her to have had a child so soon. After our European tour, however, we discovered to our joy that we could expect to become parents towards the end of July. I wrote off at the end of October to my father in Bombay, who had of course already learnt about it from our family doctor S. K. Shangloo and was cross that I hadn't informed him earlier. He immediately cabled Beryl Stileman to come out to be with Asha during her confinement, which was planned to be in Bombay, where maternity care was the best available in India. We flew down to Bombay early in December, and the great gynaecologist Dr V. N. Shirodkar examined

Asha. It was decided that I would leave her in my father's flat and return a couple of weeks before the baby was expected.

In 1956 my mother had come to Kashmir from Kangra to be near us, and lived there in the Chashmashahi House, where she and my father had been married around 1927. In Jammu she lived in Amar Mahal, where she had spent so many years in the pre-partition days. She was delighted to be back, and I was also happy that she was with us again. However, the Indian mother-in-law syndrome is fairly durable, and although there was never any open conflict there did develop some tension as far as Asha was concerned. Naturally when my mother discovered that Asha was expecting a baby she was thrilled, as her yearning for a grandson even exceeded my father's. She had desperately hoped that despite the earlier estrangement my father would ask her to come to Bombay for the confinement, but he was not a man to forget and forgive easily, and having been upset at her leaving after our wedding he did not do so. Although this came as a sharp disappointment to my mother, it was perhaps just as well because with both my parents in the Bombay flat the tension for Asha would have been unbearable. As it was, she had to spend several months there without me. My father, though extremely fond of her, was a moody and difficult person, and it was only Beryl's presence that helped to relieve the strain. She and Asha used to go for long walks around the Mahalakshmi race course every morning. Meanwhile my father started meticulously buying all that was needed for the baby, hiring a nurse, and booking two rooms in the Breach Candy Nursing Home on the sea, one for Asha and the baby and the other for visitors including himself. He also asked me to send him the papers connected with my birth, evidently because he wanted to plan what sort of celebrations were to be held if a son was born.

I returned to Jammu after leaving Asha in Bombay, and resumed my studies for the M.A. (Final) examination that was due in April. In the second week of February I went back to Bombay. By then Asha's pregnancy was advanced, and this being my first experience of ever really seeing a pregnant woman at close quarters I found it rather alarming. However, she had been looked after very well by my father, Beryl and Dr Shirodkar, and all one could do was to wait. Irritatingly enough I went down with an attack of para-typhoid (psychosomatic?) and was in bed when Asha's pains began on 23 February. She and Beryl went off to Breach Candy, followed by my father and the staff. I was left behind in the flat.

Early on the 24th morning a baby girl was born, and though she was to become the apple of our eye I am constrained to record that my father was deeply disappointed. Beryl reported that when the nurse came out of the delivery room and announced that it was a girl he muttered something under his breath that sounded suspiciously like 'damn'. I think he had been banking on a boy because he had wanted to transfer to him a good deal of his jewellery and personal fortune. However, he was careful to conceal his disappointment, and when he saw Asha he congratulated her warmly and gave her a present. My mother's reaction when my telegram reached her in Kangra was even more dramatic. She fainted in the arms of her maidservants and, on recovering somewhat reluctantly, refused to take her normal food for almost a week.

This curious reaction brought home to me again the strangeness of some of our responses. On the one hand, the woman is given pride of place in the Hindu pantheon, being invariably worshipped before the divine spouse (*Sita*–Ram, *Radha*–Krishna, *Lakshmi*–Narayan, *Gauri*–Shankar), and on the other, the birth of a daughter is considered to be a stroke of poor luck, and an occasion for commiseration. If baby girls were not born the human race would soon disappear; yet in these responses reason often takes the back seat. In any case my father kept a stiff upper lip, and as far as I was concerned I was delighted with the baby. She looked so tiny and frail to begin with that I was almost afraid to carry her. But she marked an important event in our lives, cementing the relationship between Asha and me and becoming a source of great joy to both of us. She was the first girl born in our immediate family for almost a century. We called her, appropriately enough, Jyoti, a lamp of brightness and cheer.

Chapter 15

General elections to the State Legislature were held in the latter half of March 1957. Unfortunately a trend was set in the valley which remained firmly established for several years to come. Of the 43 seats, 35 went uncontested to the National Conference, thus virtually pre-empting the results in the 75-member Assembly. This was because the nomination papers of opposition candidates were in several cases rejected on flimsy grounds. Bakshi Ghulam Mohammad, with all his administrative ability, was an expert in 'managing' elections. In the final result the Praja Parishad in Jammu led by Pandit Prem Nath Dogra was able to win only five seats, and the rest went to the National Conference, ensuring another five-year term for the Bakshi Ministry. The pro-Abdullah forces were kept entirely out of the Assembly, and did not even contest the elections.

Meanwhile, I had been working hard for the second part of the M.A., and in April went to Delhi to sit for the final examination. As it turned out, I did my essays even better than I had expected. When the results were announced later in the year I found that I had averaged 70 per cent for the eight papers, and had set up a Delhi University record for M.A. Political Science which still stands. It wasn't really fair competition, because while the other examinees had been only students I had for many years been an active participant in political affairs and constitution-making. Nonetheless, I was pleased with the result, as I had put in a good deal of sustained labour despite my other preoccupations.

By the time we moved up to Srinagar for the summer the second general elections had been held, and Jawaharlal Nehru had been returned to power. There was never any doubt that the Indian National Congress under his leadership would sweep the polls, but the surprising thing was that the Congress was returned with less than 50 per cent of the total polled votes, a pattern that continued for many years. Senior colleagues who were with him earlier, including Sardar Patel, had passed away; so Govind Ballabh Pant had been brought in from Uttar

Pradesh to handle the crucial Ministry of Home Affairs and became an important figure in the Government of India.

Indira Gandhi, though not yet occupying any official position, was beginning to play a more significant role in political affairs. Having known her for several years as hostess in her father's house, I saw it as a welcome development because I could discuss problems of Jammu and Kashmir with her in much greater detail than with Jawaharlal. She was an excellent listener, and though generally she did not say very much she absorbed information and ideas well and was becoming increasingly influential with her father. At that time few would have been able to predict her spectacular future political career, but already she was developing clear and strong views on various issues, and her unique vantage point gave her a nationwide appeal that none of the post-Jawaharlal generation of political leaders could match. Also, in refreshing contrast with the old guard of the Congress leadership, she had a modern and forward-looking attitude to national problems.

She was in Srinagar that summer, and I discussed with her a question that was beginning to assume disturbing dimensions. While Bakshi's pro-India sentiments and actions were welcome, his administration was becoming more corrupt and the calibre of his Ministers was generally poor. In those days blatant corruption was still looked upon with disfavour, and enough of the idealism of the freedom movement remained for a corrupt Chief Minister to be a major embarrassment. Despite my support for Bakshi I did not approve of this, and asked Indira Gandhi to convey to Jawaharlal my request that he and Pantji should intervene to improve the situation. With regard to the new Cabinet I said that though Bakshi may be given a free hand in its formation, the importance of removing the abuses in his administration and party should be impressed upon him. I also urged the appointment of a Chief Secretary from the Centre, the speeding up of the integration of State services agreed to earlier, and the revival of the post of a full-time Secretary for Kashmir Affairs in the Government of India so that a senior civil servant could monitor continuously the various political, financial and administrative dimensions involved in handling this most sensitive of Indian states.

Despite Indira Gandhi's importance, Jawaharlal remained

the fountainhead of political power. His prestige in India and abroad was unassailable, and despite the heavy pressure of administrative and parliamentary responsibilities he retained the wider vision and depth of perception that sets a world statesman apart from his contemporaries. Always a man who combined action with deep thought, and who wrote with rare charm and elegance, Jawaharlal took to writing fortnightly letters to Chief Ministers setting down his thoughts on a wide spectrum of national and international problems, probing the depths of his own mind and the dim corridors of history to come up with fresh insights into the events of the moment. Indeed this capacity to set India's problems in the broader perspective of world history was one of his striking achievements.

From 1956 onwards he started sending me regularly copies of his fortnightly letters, which were an endless source of interest and inspiration. They showed how, despite his strong views, Jawaharlal steadily moved towards a rounded and more mature attitude towards life and its varied problems. In particular, I was struck by a letter he wrote on 25 October 1957 after returning from a visit to Japan, Hong Kong and Burma. In this he said, *inter alia*:

> We live in a country where a multitude of conflicts bear us down. Some people imagine that it is only through conflict that the right can triumph. It may be so. But the question is whether the conflict should be based on hatred and violence, or on an adherence to one's principles and, at the same time, always offering the hand of friendship. . . . And so, I come to the conclusion that it is more important to adopt the right way, to pursue the right means, than even to have the right objectives, important as that is. No method and no way which is bound up with the creation of hatred and conflict and which bases itself on violence, can ever yield right results, however good the objective. That I think should apply to our national and international policies as well as to our political and economic approaches. If it was once clear that our approach was going to be devoid of hatred and the spirit of conflict, and discarded violence, then it does not matter very much what path we pursue. The errors we make will be corrected.

Mahatma Gandhi knew exactly what he was saying when, many years earlier in the course of the freedom struggle when Jawaharlal often differed sharply with him, he had said, 'Do not worry, when I am gone he will speak my language!' Always keenly aware of the spirit of the times, Jawaharlal was one of the first statesmen of the post-colonial era to grasp the sweeping changes that were engulfing the world. In the same letter he wrote: 'A new age and, indeed, a new world is gradually unrolling itself before our eyes. The tremendous advances of science and technology are far more revolutionary than so-called revolutions in the political and economic spheres. All our past thinking is out of date, even as all past methods of warfare are completely outmoded today. Yet we think in the old way and try to solve problems of tomorrow by the methods of yesterday which are dead and gone. We shall have to think afresh not only in India but in the rest of the world.'

Looking through the files I find that on this letter I had recorded the following note: 'This letter reveals beautifully the present political philosophy of Panditji. He seems to be mellowing and advancing spiritually, indeed to be developing into something of a sage.'

The Indian Philosophical Congress held its thirty-second annual session in Srinagar in 1957, and I was asked to inaugurate it. I had always been attracted to philosophy, not as some sort of mental gymnastics but in the more traditional Hindu sense of representing a quest for the *para-vidya*, the higher knowledge, through which alone man can move towards the fulfilment of his individual and collective destiny. In my speech I dwelt upon the unique philosophical and multi-religious tradition of Kashmir, which successively nurtured the Vedic, Buddhist, Shaiva and Sufi traditions and produced such outstanding thinkers as Acharya Abhinavagupta and mystics such as Yogin Lalleshwari and Sheikh Nooruddin. I then touched on one of my main preoccupations, the danger caused by the sharp divergence between the exponential growth in our technological capacity for creating weapons of mass-destruction and the atrophy of our moral and spiritual faculties. 'It is indeed a tragedy,' I said, 'that while we have been able to split the atom and to explore the infinitely complex structure of matter; to travel faster than the speed of sound; to explore the depths of the earth and the ocean and the infinite reaches of space, we have not been able to study our own being or to grasp the true nature of the human personality.'

My own attempts to tread the inner path continued. Music remained a major interest and I had restarted my singing, this time under the guidance of the leading musician in Jammu, Pandit Uma Dutt Sharma. Although he never became famous as a performer, he was a remarkable reservoir of classical music in every conceivable *raga* and *tala*, and I greatly expanded my repertoire by learning numerous items from him. I only regret that we did not get him to record all that he knew, which included beautiful compositions in popular as well as rare *ragas*. His young son, Shiv Kumar, was a brilliant tabla-player, and when he was only twelve I sent him to participate in an important music conference organized every year in Bombay. He made a tremendous impression, but soon shifted to the traditional Kashmiri instrument, the Santoor, which had never before been used for Hindustani classical music. Shiv excelled in it, and is now recognized as the world's outstanding exponent of the Santoor.

It was when immersed in one of the marvellous classical melodies that I would discover hitherto submerged elements in my psyche, almost as if I was a being apart listening to my own voice. Many of Pandit Uma Dutt's pieces were from the Agra–Mathura school dedicated to the worship of Sri Krishna, and, once again, the enigmatic figure of the blue god began to take a predominant place in my worship of the divine. That music is a potent source of spiritual inspiration is testified to by almost all the world's great religions. Even in Islam, which is generally not favourably disposed towards music, the Sufis sang and danced in ecstasy as their dry orthodoxy dissolved in the crucible of divine realization.

Apart from music, I was continuing my daily puja, and a new dimension was added through a chance contact. My essays on a pilgrimage to the famous Amarnath shrine in 1953 had been published in the *Hindustan Times*, and evoked several letters from people I had not known personally. One of these, a Ram Peare, wrote to me saying that the articles revealed that I was ready for spiritual guidance and suggesting that a person called Raushan Nath could perhaps be of help. I responded to the letter, and after a few months Raushan Nath, or Nathji as he is called among his circle, met me in Jammu. He was then of middle age, and his technique lay in a high-pitched and remarkably evocative chanting of the single syllable 'Aum'. This is *par excellence* the 'sacred word' of Hinduism, and receives special mention in several Upanishads and the *Bhagavad Gita*. Interestingly, it is clearly linked with the 'Amen' of Christianity and the 'Amin' of Islam, proving again that beneath the jungle of differing beliefs and rituals there is a deep underlying unity among the great world religions.

So far I had known of 'Aum', but had not realized that it could be used as a potent instrument of psychic development. Nathji would sit on the floor in my puja room and start chanting the mantra, gradually increasing in intensity until it was like a powerful jet engine ready for take-off. I would sit quietly, joining in if I felt so inclined. This contact with Nathji, which included extensive correspondence, lasted several years. Occasionally I had a curious sensation as if a fluid substance was moving at the base of my spine, and on one occasion it jumped half way up my spinal column. This was evidently a manifestation of the *Kundalini* or serpent-power believed by yogis to reside coiled at the base of the spine.

Sir John Woodruffe (Arthur Avalon) several decades earlier, Pandit Gopi Krishna in more recent times and some others have written at length about the *Kundalini*, its arousal and the powers it brings in its wake. The basic work on this is the celebrated *Yoga Sutras* of Patanjali, written many centuries ago. While all these texts say that the arousal of the *Kundalini* comes as a result of prolonged discipline, including sexual abstinence, and is accompanied by ecstatic states of consciousness, I must confess that I had no such experience. Perhaps this was because I was leading a normal married life and had not undergone any penance or other discipline. However, I *did* distinctly feel the movement in the spine, rather like the column of mercury jumping up and then gradually settling down in a blood-pressure machine.

At about this time I came across another person who was clearly a man of spiritual achievement. He was Swami Ashokananda, who had, after many years of penance, set up an ashram dedicated to Sri Ramakrishna at a beautiful spot known as Nagdandi just beyond Achabal at the western end of the

Kashmir valley. He had a gentle mien and was a man of deep learning. I began visiting him occasionally during the Kashmir season, and it was a privilege to sit in his sylvan retreat and listen to his talk of the spiritual quest in a soft Bengali voice, barely audible against the gentle laughter of the brook that flowed nearby.

The contact with Nathji and Swami Ashokananda helped to bring to the surface my desire for spiritual progress. I had begun writing poetry, and my early poems reflected this craving. I was in search of a friend, the divine friend, who would take me by the hand and lead me into new pastures. With my curious childhood and the plunge into public life at the age of eighteen I had missed an important ingredient, real friendship. My cousin Nasib remained a close confidant, and had begun looking after my estates in Kashmir. Asha, of course, had shared my life since she was thirteen and I nineteen, and we had virtually grown up together. After Jyoti's birth there was a period which called for some mutual adjustment, but we managed this quite successfully. Nonetheless, I felt an inner void.

The temple was built almost at the top of the mountain. Our car had brought me to the village of Panuanaula seventeen miles beyond Almora, and from there I had walked up for about a thousand feet through the forest aided by a stick and a servant, until I finally arrived at an idyllic spot nestling in the mountains about seven thousand feet above sea level. The temple, with its grey dome, stood in absolute silence, and the area appeared to be quite deserted except for the chirping of birds and rustling of leaves. After a search I located a gardener who told me that Swamiji and his party were out for their evening walk. So I sat on a stone bench and, as I awaited their return, I thought about the man I had come to meet.

I had first read about him in Dilip Kumar Roy's book *Among the Great*, in which he had recorded five interviews with eminent philosophers. The one on Sri Aurobindo was particularly interesting, and not only did it whet my appetite to learn more about that extraordinary man, it also mentioned an Englishman Ronald Nixon, later renamed Sri Krishnaprem, who had come to India in search of enlightenment. On my writing to him, Dilipda sent me a book entitled *Sri Aurobindo came to me*, which contained a chapter on Sri Krishnaprem. I found him absolutely fascinating, and as my interest in Sri Krishnaprem grew I decided to seek him out. Dr K. M. Munshi, a man of deep scholarship, was then Governor of Uttar Pradesh, and I secured from him Sri Krishnaprem's address. Thereafter I started corresponding with him; his letters were answered on his behalf by his disciple Sri Madhava Ashish. It was after about a year's correspondence that I had decided to make a trip to Almora on 8 December 1958.

Ronald Nixon had been a fighter pilot in the Royal Air Force during the First World War. After the war he had come to India in the early twenties to teach English at the Lucknow University. By one of those strange quirks of destiny which can be attributed either to chance or to divine dispensation, depending on the predilection of the attributer, he found his guru in the saintly wife of the Vice-Chancellor of that very University, Professor Chakravarti. Sri Yashoda Mai soon thereafter gave up the householder's life and eventually retired to the Himalayas. Sri Krishnaprem resigned his Lucknow job and for a while moved to Banaras at less than half the salary. Thereafter he also became a *sanyasi*, donned the ochre robe and followed his guru to Mirtola where in the early thirties they established a Vaishnava temple dedicated to Sri Krishna. After several years they were joined by two other Englishmen, Sri Haridas and Sri Madhava Ashish, and later by several Indian disciples.

Sri Krishnaprem was a remarkable man, with a scintillating mind and a powerful thrust towards the divine. A brilliant Cambridge graduate, he became a serious student of Sanskrit and Pali, and could speak fluent Bengali and Hindi. He had written several books which I had read before meeting him, notably a series of articles entitled *Initiation into Yoga*, a luminous commentary called *The Yoga of the Bhagawadgita* and another, *The Yoga of the Kathopanishad*. What was particularly impressive about his writings was the manner in which he drew on the Western mystical tradition—Plotinus, Hermes Trismegistus, Meister Ekhart—to show that the spiritual quest was a universal one and not confined to any particular religious tradition.

His penetrating insight was matched only by his total devotion to Sri Krishna, 'Thakur' as he called him. Dilipda was responsible for his visit to Sri Aurobindo in Pondicherry, and in a letter to him the great sage himself had written of Sri Krishnaprem: 'You know very well the value I have always put upon his insight into spiritual things, the brilliance and accuracy of his thought and vision and his expression of them. I think I described it once as *pashyanti vak*.' No wonder, then, that I awaited his return from the evening walk with keen expectation, deepened by the absolute serenity of the temple and the spiritual aura of the Himalayas in which it nestled.

After about half an hour I heard voices from afar, and then the small party emerged. Sri Madhava Ashish, a tall, handsome Englishman looking like Gregory Peck in ochre, strode forward and greeted me. Behind him was Sri Krishnaprem, almost as tall, with short cropped grey hair, bespectacled, and with the bluest eyes I have ever seen. I bent to touch his feet and he embraced me with such affection as though we had known each other all our lives. I can still feel the coarse texture of his robe and the strength of his arms as I looked into those flashing eyes which seemed to radiate love and power.

I was lodged in a tiny cottage near the temple which had belonged to Moti Rani, Yashoda Mai's younger daughter, who, despite her chronic ill health, had accompanied them to Mirtola and had died a few years earlier. *Aarati*, the evening prayer at the temple, was at seven. I stood as if in a trance while Sri Krishnaprem sang *bhajans* and did the *kirtan* in flawless Hindi, accompanied by Sri Madhava Ashish on a small double-sided drum. When the lamps were lit the third person there, a Rabindranath Banerjee, beat on the metal gong and the clear mountain air resounded with the sound of worship. I had never come across anything so simple and moving before. I just sat on the parapet, gazing at a distant star.

The temple was ingeniously constructed with a number of small rooms both on the ground and first floors, which could be kept warm during the long Kumaon winters. Sri Krishnaprem's room was on the first floor leading directly into the shrine, but visitors approached it through a flight of stone steps from the back. Here we assembled after *aarati*; Sri Krishnaprem sat at one corner next to the fire, cooking delicious potatoes and *laddoos*, while Sri Madhava Ashish sat opposite him. I was given a seat on the cot, and we talked deep into the night. It was amazing how alert and nimble their minds were, and what a tremendous sense of humour they had. Contrary to the popular conception that gurus are sombre, rather boring fanatics, here was a pair full of fun and laughter. In one of his books Sri Krishnaprem says: 'Laughter was given by the Gods to man and it was one of their choicest gifts. No animal can laugh nor does it need to, since it lives in the harmony of the purely instinctive life. It is only Man whose possession of an ego introduces stresses and strains which cannot be avoided and for the healing of which, therefore, the Gods gave him this supreme gift. Time and again it will save us when otherwise all would be lost. He who cannot laugh, he whose devotions are too serious for the healing waves of laughter, had better look out; there are breakers ahead.' Dilipda sent the quotation to Sri Aurobindo for his comments. Pat came the reply next morning: 'Sense of humour? It is the salt of existence. Without it the world would have got utterly out of balance—it is unbalanced enough already—and rushed to blazes long ago.'

Nonetheless, though a sense of humour helps greatly, the spiritual quest is no laughing matter. For the three days I spent there we had extended discussions about my life-situation. I gained a number of valuable insights which have since stood me in good stead. He talked of how every action has the elements of light and shade, and how the inner price has to be paid for every apparent outer advance. It is the brightest light that casts the deepest shadow, and he stressed the importance of coming to terms with one's inner urges rather than brushing them under the mental carpet where they fester and finally erupt in unhealthy psychic manifestations.

At one point he said, 'I am not prepared to advise you on what to do or what not to do, but I will say one thing—do not lose the light that is within you. You have it and it is more precious than your intelligence, fame or wealth. It is the *only* thing that will last; don't lose it. Follow what seems to you to be the highest. You will never regret it.' He also stressed the importance of integrating the two halves of the human psyche—the male and the female, the emotional and the intellectual. Without this integration man remains a partial being constantly at war with himself, unable to move onward across what the *Katha Upanishad* calls the razor-edged path.

During the conversation Sri Madhava Ashish would join in from time to time in such an effortless manner that it appeared almost as if I was talking to a single mind divided into two bodies. The relationship between the guru and his disciple—'Gopalda' and 'Ashishda' as they were called—was unique. Sri Krishnaprem's words came from the depth of great spiritual achievement, the vision of a true seer. When he spoke of Krishna, his beloved, the eternal lover of all beings, his eyes would glow with a strange radiance and his whole body seemed vibrant with inner joy. Never will I forget the glory of his presence when, along with Ashishda, the three of us would sit by the glowing embers in that little room in Mirtola, talking deep into the long winter nights. We talked of everything—politics, people, relationships, books, dreams. To every topic they would bring a fresh approach, constantly linking outer activity with inner aspiration. They looked upon life as a series of concentric circles, each covering different areas of activity but all centred in the self. Here were the great teachings of the Upanishads brought alive by a pair of vibrant Englishmen born and raised ten thousand miles away from India. Keen intellectual perception combined with deep emotional empathy enabled them to analyse my life-situation in a manner no one else had ever done, and in the process gave me a deeper awareness of my inner being. And even more eloquent than their words were their silences, when their eyes would lock into a flow of power so tangible that I hardly dared breathe lest it be disturbed.

Chapter 16

The year 1959 saw the tragic denouement in Tibet which shattered the Sino-Indian friendship so laboriously nurtured by Jawaharlal Nehru and led to the debacle of 1962, a shock from which Jawaharlal never really recovered. Perhaps it was his idealism; perhaps inaccurate reportage from our Ambassador in Peking, K. M. Panikkar, in whose insight into history Jawaharlal had great faith (he had earlier recommended *Asia and Western Dominance* as compulsory reading for senior diplomats and Ministers); perhaps it was the growing influence of the ebullient but erratic Krishna Menon; perhaps it was simply the geo-political realities of Central Asia. Whatever it was, the Tibet fiasco was the first major setback to Jawaharlal and to our hopes of a smooth upward march for the Indian nation.

For us, in particular, the Chinese subjugation of Tibet came as a rude shock. My intrepid ancestor, Maharaja Gulab Singh, and his band of remarkable generals led by Wazir Zorawar Singh, who must rank in world military history as one of the most extraordinary commanders of all times, had in the nineteenth century launched the great Central Asian campaigns that had resulted in the consolidation of India's northern frontiers and added thousands of square kilometres to its territory. After conquering Ladakh, Zorawar Singh was in fact marching on Lhasa when a particularly severe winter halted his advance and resulted in the disaster at Rupkund. Whereas other groups in India with proud martial traditions fought bravely down through the centuries, it is the Dogras alone who had the unique privilege of expanding India's national frontiers into the heart of Central Asia.

The tragedy of Tibet was felt with particular impact in the State because Ladakh was culturally very similar, and there were many religious and family bonds between that region and Tibet. About five hundred Ladakhi monks were under training in Lhasa when Chinese troops took over, including four incarnate lamas (Kushaks). Kushak Bakula flew down to Delhi to convey to the Prime Minister the apprehensions of the people of Ladakh about their welfare, as well as about the safety of the Dalai Lama, who was held in the deepest reverence by all Lamaistic Buddhists. Having met the Dalai Lama when he visited India a few years earlier, I too was deeply concerned for his safety. The dramatic disappearance of the Dalai Lama and his close followers from Lhasa, their escape under a thick blanket of cloud which lasted several days until they reached Tawang, had all the elements of a classic adventure. But while it was a relief that he escaped unharmed, the destruction of the ancient Tibetan civilization by the Chinese was a source of great anguish. Whatever the military and geographical compulsions, many Indians felt that Jawaharlal took too negative and passive a view of the whole matter. Had we moved immediately after Independence, when China was still embroiled in its terrible civil war, to ensure for Tibet the status of an independent country with a seat in the United Nations, the whole situation may well have been different. But, despite obvious difficulties, it seems that there was inadequate appreciation in those days how important an independent Tibet was for us, which resulted in the debacle of 1962 and gravely affected India's national interest and prestige.

While relations with one mighty neighbour took a sharp nosedive, those with the others continued to improve. An official delegation from the Soviet Union led by A. A. Andriyev and N. A. Mukhitdinov visited Srinagar in March and reiterated the stand taken by Khrushchev and Bulganin three years earlier that the State was an integral part of India. At my suggestion, they broke journey on their return at Jammu, where they were given a warm reception. Plans for my visit to the Soviet Union had meanwhile been settled, and on an official invitation from the Soviet government Asha and I, accompanied by Professor P. N. Chaku, paid a three-week visit to the USSR, from 23 April to 15 May 1959.

The Russians looked after us with their well-known hospitality. Our first stop was in Tashkent, which we reached on the Air-India Super Constellation 'The Rani of Ayodhya' after overflying Lahore, Ghazni and Kabul. Thereafter we continued the long flight to Moscow, where we were greeted by a Secretary of the Presidium of the Supreme Soviet, the Minister for Higher Education and a battery of photographers. By the time we reached the charming wood-panelled bungalow on 13 Alexei Tolstoy Street where we were to stay it was quite late. During the next three weeks we went through a strenuous programme in Moscow and visited Leningrad with its great Hermitage Museum; Stalingrad, later renamed Volgograd after the de-Stalinization campaign had gathered momentum; Sochi, a Black Sea Resort with impressive holiday homes for workers; Tblisi, the capital of the Georgian Republic; Tashkent, the capital of Uzbekistan; Samarkhand with its startlingly turquoise monuments built by Timur and finally Dushambe (then Stalinabad), the picturesque capital of Tajikistan. Our hosts had put a special plane at our disposal, so we were able to travel in comfort.

The visit was a most instructive experience. I was impressed by the warmth of the Soviet people, and the depth of their friendly sentiments towards India. Jawaharlal's policy of non-alignment had at last been appreciated by the Soviets, and from Khrushchev down to the ordinary citizen I found a sincere liking for Indians and high regard for India as a great and ancient nation. In several universities Sanskrit, Hindi and other Indian languages were being taught, and during our visits to the 'Pioneer Palaces', the ubiquitous children's movement that is such a prominent feature of Soviet life, we met many children who spoke to us in Hindi. At one of these visits I responded to their cultural show by singing a lilting Dogri folk song to which they all clapped in unison.

While closely geared to the official Marxist-Leninist philosophy, the Soviet educational system seemed to have a certain purposefulness and clarity of objectives which is sadly missing in India. There is no aimless drift from school to college. Only a small percentage of secondary school students, selected on the basis of carefully planned future requirements as well as aptitude tests, are allowed to enter institutions of higher education, and they are assured of employment after completing their courses. This is in sharp contrast to the appalling waste of talent and human resources that is so blatant a feature of Indian education. The total dominance of all aspects of life by the Communist Party, however, was evident. While the forms of parliamentary democracy were there—elections, secret ballots, legislatures—in practice the whole system is run by the party and there is only one candidate for each electoral seat. Whenever I gently hinted at the contradiction involved in such 'elections', I was greeted by rather embarrassed evasions from our friendly hosts.

Two aspects of the visit made a particular impact on me. With my interest in religion, I had specially asked to visit a church and a seminary. Several places were suggested, and we ultimately chose to visit Zagorsk, a small town about seventy kilometres outside Moscow which is an important centre of the Russian Orthodox Church. We went straight to the great complex of buildings belonging to the monastery, which dominates the town. As our visit was entirely unscheduled it took our hosts a few minutes to contact the local monks. A young monk, with a beard and a beautiful face, finally came out to greet us and showed us around. We first went to the Church, and as it was Easter Eve the morning service was well attended. The Church was full of worshippers, mostly elderly women, who kissed the image of Jesus lying in a small sarcophagus. My wife and I lined up with the worshippers and bowed to the relics as we passed, while our interpreters looked on slightly embarrassed. Later we were taken to meet the Vice-Patriarch, who conducted us around the Theological Institute with its museum containing a small but fine collection of religious art. Evidently this was a training centre for monks, and seemed to be supported largely by offerings from believers. Atheism is a central and essential tenet for all members of the Communist Party in the USSR, as Khrushchev himself confirmed to me in response to my query, but it seemed that religious beliefs were allowed to function through the churches, though these were strictly regulated by the State.

That evening we attended the Easter service at the great Yelohovsky Cathedral in Moscow, presided over by the Grand Patriarch Alexei of the Russian Orthodox Church. The huge cathedral was packed to capacity with devotees, many of whom were repeatedly crossing themselves while tears streamed down their faces. The chanting of the ritual was in Old Slavonic, which sounded very like Vedic Sanskrit, and the waving of the lighted candles and burning of incense to the accompaniment of choral singing was reminiscent of a Hindu *aarati*. Priests with flowing beards and red brocade vestments performed the solemn ritual. At quarter to twelve the Patriarch asked for fifteen minutes' silence and entered the *sanctum sanctorum*, emerging at the stroke of midnight carrying three lighted candles in a silver stand and swinging an incense burner. 'Christ has arisen, Christ has arisen,' he intoned in his powerful voice, to which the congregation responded, 'Verily He is Resurrected.' Then the Patriarch led the procession thrice around the cathedral, like the Hindu triple *parikrama*, where the milling crowd jostled to get a glimpse of the candelabra. It was a moving and memorable experience, this deep symbolic festival of spiritual resurgence in the land of Marxism-Leninism.

The other great experience on this trip was the glorious Russian ballet, which happily continued to receive full State patronage even after the advent of communism. We saw *Swan Lake* at the magnificent Bolshoi theatre in Moscow, with Maya Plesetskaya dancing the twin Odette/Odile role like some ethereal creature. In Leningrad we saw *Spartak*, based on the life of the leader of the great slave revolt against the Roman tyranny, and also *Don Quixote* and *Othello* and, in Tashkent, an opera called *Din-Aram* based on a famous poem by Ali Sher Navai. Both the hero and the heroine were Kashmiri, and the scene in Kashmir had a set combining the architectural features of the Shah Hamadan mosque in Srinagar and the Raghunath temple in Jammu!

Politically, the highlight of our visit was a private lunch hosted by Khrushchev at the Kremlin. After a forty-minute interview in his office he walked with us to another part of the Kremlin for lunch. We had a drink in what he told us was the ante-chamber of Catherine the Great, and lunched in the adjoining room. The intimate party, apart from Khrushchev and ourselves, included Andrei Gromyko and his wife, India's Ambassador K. P. S. Menon and his wife, and Mukhitdinov, the Uzbek member of the Party Presidium who had visited Srinagar earlier that year. Khrushchev was a jovial host and talked a great deal. He said that he was glad to see both of us again looking so well, and remarked that he thought my wife's eyes had grown darker since he had last met us. In turn, I observed him closely. He appeared to be slightly less exuberant than during his visit to India, but spoke with added assurance and power. At the end of the lunch, after endless toasts in the

Russian fashion, he made a speech that was reported in some detail next morning in the *Pravda*. He made friendly references to India and Indo-Soviet relations. Speaking about communism he said: 'Some people abroad dread the word communism. Communism is a great idea and it is not abstract. Its essence is material goods; food for the people, high level of the economy of the country, plants, factories; a wide network of high and secondary schools and cultural institutions.' Before concluding he observed: 'It is with pleasure that we often recall our visit to India and the hospitality accorded to us by the great Indian people, their government and in particular by the people of Kashmir and personally by you.'

I sent the Prime Minister a detailed note on my visit to the Soviet Union, and he circulated it to members of his Cabinet for their information. I also kept in touch with him regarding political developments in Jammu and Kashmir and broader national issues. I was getting increasingly interested in national politics, and my talks with Sri Krishnaprem had revolved to a considerable extent around whether or not I should join them. This was in the late fifties, and at the time he had advised against it because he felt that I still had important work to do where I was. However, I realized that although Kashmir was important it was only a small part of our great nation, and its problems could not in the final analysis be solved in isolation. I had no intention of spending the rest of my life as *Sadar-i-Riyasat*, which was one of the reasons why I had not favoured the suggestion made in 1952 that I should demand a life-term before accepting the office.

On 13 July 1958 Jawaharlal had appended a seven-page note with his fortnightly letter to Chief Ministers in which he expressed his views, 'loud thinking' as he termed it, on a number of major problems facing the country and the world. It was a brilliant piece; the earlier vision of the *Discovery of India* had been enriched by ten years of active administration as leader of the world's largest democracy. He discussed the merits as well as weaknesses of both capitalism and communism, and seemed to seek a third ground which would be a creative synthesis between socialism and Vedanta. In fact, he mentioned the

Vedantic concept several times in the note, and ended by saying: 'In considering these economic aspects of our problems, we have always to remember the basic approach of peaceful means; and perhaps we might also keep in view the old Vedantic ideal of the life force which is the inner base of everything that exists.'

After reading this note I decided to write to Jawaharlal. 'Along with your letter dated 13 July,' I wrote to him on 8 August, 'you attached a note in which you indulged in some "loud thinking" upon the problems that face us. The result was most thought-provoking. Belonging as I do to a much younger generation [I had just turned 27] I thought you might be interested in some random reflections that your letter induced in me, which I have set down in the enclosed note.' I reproduce below the note *in extenso*, because it reflects my thinking at that time and shows that several issues—population control, nuclear disaster, corruption in public life and spiritual renaissance—which have remained important for me were already clearly formulated long before I plunged into national politics.

1. At the outset it may be repeated, because it is not always realized by older people, that ours is what may be termed the post-Independence generation in that our active political thinking and activity began after India had become free. Thus even the great freedom struggle, in which you were privileged to play so distinguished a role, has not really got a deep personal significance for us and we judge men and ideas by their post- rather than pre-Independence performances. This is not to say that we were indifferent to the freedom movement. On the contrary even as teenagers we were thrilled to the core by the great nationalist upsurge that swept the British out of the country. I still remember the feverish excitement when I first (and last) met Gandhiji here in 1947, and again when for the first time you came into my sick-room at Jammu later the same year. I recall how avidly I used to read your books in my school-days, but surreptitiously for fear that my father would see me and discover that I was fast becoming a youthful revolutionary, as indeed I was! But my point is that by the time we reached maturity the British had already departed, and the freedom struggle is thus not an integral part of our emotional make-up as it must be of older generations. Even Gandhiji is for us a somewhat shadowy figure, very great of course but nevertheless belonging to the past rather than to the present or future.

2. After centuries of servitude we became free, and Mother India, though grievously amputated and bleeding, awoke to a new-found strength and vigour. India's freedom was an event of historic significance, symbolizing and heralding as it did the re-awakening of Asia after a torpor of ages. A thrill of pride shot through us as we took our rightful place among the great nations of the world and, under your inspiring leadership, began the Herculean task of national reconstruction. In the sphere of foreign policy we have made a creditable contribution towards the maintenance of world peace and a lessening of international tensions, and your advocacy of Panch-Sheel has given the world a new philosophy of peaceful coexistence which will have to be accepted if mankind is to survive in this nuclear age.

3. Internally also we embarked upon an attempt at planned economic development, and in this first decade of Independence we have made much progress on many fronts. But at the same time there is much that is unsatisfactory and indeed alarming in our present situation. The idealism that inspired public life in the half century preceding Independence appears to have largely evaporated, leaving behind a noxious residue of self-seeking opportunism. As you have so often pointed out in your speeches and writings, fissiparous tendencies seem to be growing and caste and communalism, though outlawed by our Constitution, appear to be getting more firmly entrenched. Several observers have pointed out that these forces played a much more important role in the Second than in the First General Election. Corruption has established itself at almost all levels of public life, and it is this above all that corrodes the idealism of youth. Indeed our generation, which is always the mainstay of idealism, seems to have become largely frustrated and cynical. As you point out, this is to some extent a world-phenomenon. But it is yet surprising that this should be so in India, where vast fields of constructive work have opened out before us since Independence and fresh vistas of progress lie ahead.

4. The older generation of national leaders no doubt retain their idealism, but they are necessarily fast diminishing, leaving gaps in public life which are difficult to fill. The recent demise of the Maulana is a case in point. As Sri Aurobindo wrote in 1909 on the passing away of Romesh Chandra Dutt, 'The landmarks of the past fall one by one and none rise in their place.' I am more than forty years younger than you, and between our generations lies a vast gap only occasionally lighted by outstanding inspiration and idealism. You stand as a titanic figure, forward-looking, young in mind and spirit, full of noble ideals allied with the constant endeavour to actualize them. But, with a handful of exceptions, you stand alone, though we pray that you may be spared for many more years to guide us.

5. What is the cause of this lack of idealism in our country, even among the younger generation, as is so often reflected in our universities? I for one do not feel myself competent to offer an analysis of the complex economic, social and psychological factors that operate to shape the consciousness of our youth. But I think a fair share of responsibility must be borne by our political parties whose task it is in a democracy to infuse a spirit of hope and dedication among the people, especially the younger generation. They appear to have failed in this vital duty, and almost without exception have proved incapable of evoking in our generation genuine idealism on any substantial scale.

6. In your note you mention the views of a colleague: 'In our attempts to ensure the material prosperity of the country we have not paid any attention to the spiritual element in human nature. Therefore, in order to give the individual and the nation a sense of purpose, something to live for and, if necessary, to die for, we have to revive some philosophy of life to give, in the wider sense of the word, a spiritual background to our thinking.' I feel that this is a crucial point, because it is widely admitted that the general level of national character—discipline, honesty, integrity—has seriously declined and continues to do so. This is a very serious matter because without a high level of national character the new India of our dreams will remain a mirage. It is our educational institutions and, perhaps, the Community Development organization that must spearhead this 'spiritual renaissance', if I may use the term in its broadest sense. But mere words are not enough, concrete steps are called for before it is too late. The spiritual motive has always been an important element in our past greatness—as witness the Buddha, Ashoka, Gandhi—and if our country is to achieve the pinnacles of greatness in the future she must adhere to this ideal.

7. Finally a word about our economic problems. Whether or not a spiritual motive permeates us, there is no doubt that we will be judged, and indeed our democracy itself will be judged, by the extent to which we can solve our economic problems and ensure to the people of India a reasonable minimum standard of material goods and services. We have of course achieved a good deal since Independence, but the expectations of our people always expand faster than our capacity to meet them, with the result that the gap remains and even grows. I am no economist and so I will not venture to say anything with regard to the method or content of our economic planning, but there is one matter which I feel bound to mention and that is the question of population. In almost all your letters you lay great stress upon the supreme importance of raising agricultural production and the need for a vigorous, sustained and co-ordinated drive in this

direction. This is of course extremely important and must be done. But however much we may increase our food output per acre, however much we may extend our irrigation and improve our methods of production, our huge and rapidly increasing population will inexorably swallow up all that we produce and millions will remain in economic distress. It appears that the disastrous consequences of our terrifying population growth are not really being brought home to the people. Our public leaders talk constantly on a large variety of subjects but are conspicuously silent with regard to the supreme necessity of limiting our population. It may be argued that mere words will achieve nothing unless we have the concrete means of mass birth control. This is of course true, but before any such campaign can be launched we must create the necessary climate of opinion in the nation, particularly in the rural areas. Are we making full use of all the modern means of publicity at our control to educate the nation with regard to this basic problem? We are spending hundreds of crores upon numerous development works, but are we spending enough upon family planning and birth-control on a mass scale? Even China, which at one time impatiently brushed aside the poulation problem as a 'capitalist myth' which would automatically disappear in a communist society, has now, I understand, adopted stern and active measures to limit its population growth. Unless we do the same all our plans and schemes will be gravely jeopardized and we will never succeed in lifting the mass of our people out of the morass of poverty in which they have been sunk for so many centuries. Population control is the most fundamental and important national problem facing us today, not excepting even food production. Have we a proper awareness of this, and if so, are we doing all we can to meet it?

After a few days he wrote back:

My dear Tiger,

Some time ago, I received your letter of August 8th. I read your note with great interest and, I think, profit. What appealed to me especially was a young man of your generation looking at the world as it is. I wish I could enter a little more into the mind of your generation.

Yours affectionately,
Jawaharlal Nehru

Meanwhile, political developments in the State proceeded on two lines. On the one hand, important measures for bringing Jammu and Kashmir closer to the rest of the country continued to be taken by the Bakshi government. The system whereby even Indian citizens needed a permit to enter the State was abolished from 1 April 1959, which was an important political and psychological step in the direction of closer integration. A Bill was passed amending the State Constitution, bringing our High Court judges on a par with those in the rest of the country, and also extending to the State the jurisdiction of the Indian Election Commissioner. I had been unanimously re-elected *Sadar-i-Riyasat* for another five-year term towards the end of 1957 and behind the scenes was active in encouraging Bakshi and his Ministry to move closer to the Centre. Pakistan, of course, continued to protest against all such measures.

While constitutional events moved favourably, in the political sphere troubles had begun to develop. Partly these were due to Bakshi's authoritarian manner and the widespread corruption that had grown among his relatives and party functionaries. Bakshi Abdul Rashid, the President of the National Conference, who had a bad reputation and was operating through well-known local roughnecks, was a cousin of his. This group included some other relatives of Bakshi, and began to be referred to sarcastically by his opponents as the B.B.C.—Bakshi Brothers Corporation. The dominance of this group alienated a more intellectual and ideologically-oriented section of the National Conference led by G. M. Sadiq, a veteran and widely respected leader. He had been President of the Constituent Assembly and later joined Bakshi's Cabinet, but the two men were diametrically opposed to each other in almost every way, and over the years Sadiq and his group, which included Mir Qasim, the Speaker G. R. Renzu and the brilliant D. P. Dhar, drifted away from Bakshi. This rift finally led to the resignation of the whole group and their forming the Democratic National Conference. As Head of State I had to keep in touch with both groups so that I could correctly report to Jawaharlal on the political configuration. 'D.P.' was a good friend, and kept in close touch with me even though Bakshi looked somewhat askance at our relationship. I appreciated the role played by Bakshi in strengthening the links with the Centre, but intellectually I found myself closer to Sadiq and his friends.

The breaking away of the Democratic National Conference did not in the first instance have any marked impact on the government. Bakshi's hold on the party and the administration was absolute, and at one time it seemed as though this would last for ever. But having seen the regimes both of my father and of Sheikh Abdullah collapse overnight, I had by then learnt the valuable lesson that in politics no government is ever as strong as it looks, and that unforeseen factors can and often do intervene when they are least expected. As I will relate, it was just such a factor that arose a few years later and brought the Bakshi regime to a swift end.

A more serious political development had taken place early in 1958 with the release of Sheikh Abdullah. This had been under consideration for some time, and despite our apprehensions the decision was ultimately taken by Jawaharlal and implemented by Bakshi. On his release the Sheikh entered Kashmir valley to a massive welcome, and at once launched into an angry and powerful tirade against Bakshi and the Government of India. Going back on his previous stand that the State was an integral part of India, he insisted that the accession was provisional, and that it could only be settled after the Kashmiri Muslims had exercised their 'right of self-determination'. Despite his undoubted stature and charisma among the Kashmiri masses, Sheikh Abdullah was never able to outgrow his *Kashmiriyat* and become a truly national leader.

After his release he launched a virtual crusade against Bakshi's National Conference. From the very beginning of his career he had openly used the Hazratbal Shrine in Srinagar as the venue of his political speeches after the *Juma Namaz*, the Friday prayers. It is really quite extraordinary how this blatant politicization of a holy shrine became accepted as part of life in Kashmir, whereas if any Hindu were to make a speech within the premises of a temple he would at once be dubbed communal and probably arrested on the spot. It was for Hazratbal that Sheikh Abdullah reserved his sharpest verbal barrages, and for three months he was active in whipping up political and religious-oriented mass hysteria against the Bakshi regime in particular and India in general. It was clear that he had decided to utilize the Plebiscite Front, earlier established by his chief lieutenant Mirza Afzal Beg, as his main political vehicle for capturing power. This was undoubtedly an anti-India organization, and Sheikh Abdullah's release gave it a strong shot in the arm. The whole valley was disturbed and there were frequent clashes between the Plebiscite Front and the National Conference workers. Almost all the mosques were used to whip up feelings against 'Hindu' India, and Abdullah began touring the valley extensively.

The annual 'Durbar Move' of the State capital from Jammu to Srinagar was due early in May, and it was decided to give Abdullah full freedom but to watch his activities and those of his followers carefully. For my part I was clear from the beginning that he would not reconcile himself to the post-1953 developments, but Jawaharlal and his advisers felt that he should be given a fair chance to adjust to the new realities. In the event, my analysis proved correct. After three months it was decided that his activities were compromising the national interest, and he was again taken into custody under the Preventive Detention Act on 29 April. Thus his return to office was in effect postponed for two decades, by which time the whole face of the subcontinent had changed with the break-up of Pakistan in 1971.

It must be remembered that although over a decade had elapsed since my father signed the Instrument of Accession, the Jammu and Kashmir dispute was still a prominent item on the United Nations' agenda. The reports first of the Swedish diplomat Gunnar Jarring presented to the Security Council in April 1957, and then of the American Dr Frank Graham presented in March 1958 dealt at length with the problem, and were the occasion for prolonged debates in the Security Council. It was in fact only the veto consistently exercised by the Soviet Union that prevented the passage of resolutions that India would have been quite unable to accept. The Pakistani Prime Minister, Feroz Khan Noon, who had been a member of Winston Churchill's War Cabinet along with my father, seemed to be a rare Pakistani leader who genuinely desired a fair settlement with India. In the middle of 1958 he had made friendly references to India and had said: 'war is not a solution of the Kashmir problem or any other problem.' At a time when the press and politicians in Pakistan were clamouring for a *Jehad* against India this was a courageous statement, and Jawaharlal responded readily. In September, Noon came to Delhi for talks with Jawaharlal. There was an agreement on the eastern border and an exchange of certain enclaves between East Pakistan and West Bengal was accepted although the case of Beru Bari subsequently ran into difficulties. The Nehru–Noon agreement came in for severe criticism on both sides, but it seemed to me to have been the last opportunity for a peaceful settlement.

However, this was not to be. Hardly had the ink dried on the agreement paper when Noon was ousted, the Pakistani Constitution abrogated, and a military regime installed first under General Iskander Mirza and then under Field Marshal Ayub Khan. The advent of military rule in Pakistan led directly to the Indo-Pak wars of 1965 and 1971, the break-up of Pakistan itself and the collapse of the democratic experiment as far as the Muslim-majority areas of the subcontinent (except Jammu and Kashmir) were concerned. Thereafter the Security Council's deliberations lost all real meaning, and my father's erstwhile State remained *de facto* divided on what seemed to be virtually a permanent basis. A third party to this division, unknown to us, was poised to enter the scene only a short while later.

Chapter 17

By the time the sixties began, my interest in national politics had grown considerably, and now that we had our new house in Chanakyapuri our visits to Delhi became more frequent and prolonged. Apart from seeing the Prime Minister and the President, I began meeting other national leaders regularly, including Indira Gandhi with whom I developed an easy rapport. Morarji Desai was emerging as the number two man in the government, but though he was always courteous to me I found him rigid and self-opinionated. S. K. Patil was blunt and amusing; Krishna Menon wily and mephistophelian; T. T. Krishnamachari brilliant but cynical; C. Subramaniam steady and sober; Asoka Mehta always looked a picture of misery, but had a clear and constructive mind; while Mehr Chand Khanna's weather-beaten face looked as old as the hills. Sardar Swaran Singh, with whom I conversed in Panjabi, was much shrewder than he appeared, and Humayun Kabir had a sparkling mind which I found stimulating.

Apart from Cabinet Ministers I began meeting some members of the diplomatic corps. After the election of John F. Kennedy, whose charisma and youthful idealism were such a refreshing contrast to the square and rather boring Eisenhower, John Kenneth Galbraith was appointed US Ambassador to India. We got to know him and his wife Kitty well, and they visited us both in Kashmir and in Jammu, where they once spent a weekend. He was influential with Jawaharlal Nehru, and was one of the few Americans whose opinion the Prime Minister valued. Alva Myrdal was the Swedish Ambassador and we got to know her and Gunnar Myrdal, a truly distinguished and gifted couple. However, the most stimulating person in this genre turned out to be the brilliant Arab League Representative in Delhi, Dr Clovis Maksoud. A Maronite Christian and a fierce Arab nationalist, Clovis and his first wife Rosemarie became our close friends. We saw a great deal of them, as Asha and Rosemarie also got on well. It was, strangely enough, thanks to Clovis that I met a wide spectrum of Indian intellectuals whom I might otherwise have missed. These included several radical figures, for example Aruna Asaf Ali and Jayaprakash Narayan whom I met first at Clovis' house in Chanakyapuri. Clovis brought a new dimension into my political awareness; his parties were always stimulating, and there would be protracted and often heated discussions on various national and international issues deep into the early hours of the morning. At about this time we also got to know Raj and Romesh Thapar, who had just launched *Seminar*. They remain very good friends despite Romesh's strongly held and forcefully expressed views, which do not necessarily coincide with mine. Thanks to friends like these, I began to get caught up in the intellectual excitement of the Indian political experiment, at that time still under the unchallenged spell of Jawaharlal.

Despite the growing attraction of Delhi, I did not neglect my work in Jammu and Kashmir. In 1960 there were some important developments. The Srinagar–Leh road was opened that year, marking a historic event as far as that far-flung and backward district was concerned. I paid my second visit to Ladakh, spending three days in Leh and two in Kargil, and also toured the forward areas of Tithwal along what was then the Ceasefire Line with Pakistan. A second vital road linking Udhampur with Dhar in Himachal Pradesh was also inaugurated that year, which not only opened up hitherto inaccessible areas of the Jammu region but had great military importance because Udhampur was, and remains, an important army base.

Jawaharlal Nehru, Dr Radhakrishnan, King Mahendra of Nepal and a number of other dignitaries visited Kashmir that year, and this gave us the continued opportunity to entertain and interact with an interesting range of personalities. Earlier in the year Asha and I had paid a visit to Nepal, my first ever and her first since we had been married ten years ago. I was thoroughly enchanted with Nepal, especially its famous Pashupatinath Temple, the great Buddhist shrines of Bodhnath and Swayambhunath, and the fabulous ancient cities of Patan and Bhaktpur. We happened to be there when the King inaugurated the session of the Nepali Parliament with B. P. Koirala as the Prime Minister. In the event it turned out to be its last session, because shortly thereafter 'B.P.' was dismissed, Parliament dissolved and power taken over directly by the King, thus bringing to an end the short-lived Nepali experiment with parliamentary democracy. In a way this negated the main reason for India having supported the overthrow of the Rana regime.

With Sheikh Abdullah back in detention, it had been decided after prolonged consideration and legal consultation to prosecute him and some of his colleagues on the ground that they had received illegal aid from Pakistan for anti-national activities in the State, both through individuals as well as the Plebiscite Front. This was the famous Kashmir Conspiracy Case that went on for some years before it was withdrawn on political grounds. The British lawyers, D. N. Pritt and Dingle Foot, visited Jammu in this connection, and the latter finally took up the defence. The prosecution was led by Gopal Swarup Pathak, later Vice-President, and a team of distinguished lawyers. There was incontrovertible evidence that the Plebiscite Front had received large sums of money from Pakistan and this was much later corroborated by none other than Zulfikar Ali Bhutto in his memoirs.

The rift between the ruling National Conference and the breakaway Democratic National Conference was also bridged mainly at the initiative of Jawaharlal Nehru and the intervention of Indira Gandhi. At the end of November the talks between Bakshi Ghulam Mohammad and G. M. Sadiq led to a rapprochement whereby the Democratic National Conference was wound up, and the Cabinet expanded to include Sadiq himself and three of his colleagues, G. L. Dogra, Mir Qasim and D. P. Dhar. This was a welcome development, as the rift within pro-Indian Kashmiri ranks was not in the broader national interest. However, despite the political solution there never really was a meeting of minds, and the differences between the two groups were to surface again in the future. A small rump of the Democratic National Conference continued to carry the name, led by Ram Peara Saraf and Krishen Dev Sethi, who were extreme leftists and developed links with the CPI(ML).

By now the Chinese moves in Ladakh had become a matter of national concern. They flowed from the Tibet situation, and it was amazing that despite their move into the outlying areas of Ladakh we were so slow to react. It is true that the areas involved are among the most inhospitable in the world, with no population and hardly any vegetation. Nonetheless, they were legally part of my father's State and thus of India after he signed the Instrument of Accession. Once again a major weakness of our foreign and military policy was exposed, and the northern frontier became a sort of Achilles' heel. Early in 1961 there were a series of meetings between Indian and Chinese officials on the boundary question, the report of which was published in March and debated at length in Parliament and the press. Clearly, the whole of the Indo-Chinese border had become a matter of dispute, not only in Ladakh but in the eastern sector also. Unfortunately, Jawaharlal fell between two stools; he was neither able to clinch a compromise give-and-take package, which would certainly have been attacked but which his immense prestige could have carried through, nor was he able to build up enough military strength on the northern borders to actually enforce our legal claim. The confrontation that developed between India and China was a tragedy, pitting the two most populous nations of the world with the oldest continuing cultures against each other. On the one hand was Jawaharlal Nehru, who was committed to *Panchsheel*—five principles based on the Buddhist maxims of love and compassion. On the other was Mao Tse-tung, who was committed to the Marxist-Leninist doctrine that power flows from the barrel of a gun.

Having visited Europe twice already, in 1956 and 1959, we had planned for 1961 a trip to Southeast Asia, but due to shortage of foreign exchange we had to change our plans and decided to go on a brief holiday to England. As usual, we caught our Air-India flight from Bombay, so that we could see my father, who had lived there ever since he left the State in 1949. In the fifties he also went abroad several times, and in the monsoon season would be in Poona or Bangalore for the races. But Bombay remained his home, and he lived on the top floor of Hari Bhawan at 64 Pedder Road.

Even though he was somewhat mollified by my dismissal of

Sheikh. Abdullah in 1953, our relations—which were never too warm—remained rather unsatisfactory. There seemed to be two reasons for this. The manner in which he was virtually exiled from the State in 1949 at the insistence of Sheikh Abdullah was obviously a deeply felt insult, and all the more unfair when compared with the treatment accorded to the Nizam of Hyderabad, whose forces actually fought against those of the Indian Union. My father, true to the Dogra-Rajput tradition, was a proud man and was never able to reconcile himself to what he felt was a glaring injustice. The fact that I moved into the headship of the State, under circumstances which I have described in some detail in *Heir Apparent*, must have led to a psychological resentment which could well have had an influence on his attitude towards me.

The second factor, to which I have also referred earlier, was the unsatisfactory relationship between him and my mother. I had felt, rightly or wrongly, that my mother was the more vulnerable of the two and that it was my duty to see her properly looked after. As a child my relationship with her had been much warmer, and when my parents finally separated in 1950 I did give more attention to her. After some years she came back to the State and began living near us, although in a separate house. My father was particularly angry with Thakur Nachint Chand, my elder maternal uncle, and had convinced himself that he would exploit our family. Thus, in his years in Bombay he seemed to grow increasingly bitter and alienated, although he never gave any hint of this in his letters. The fact that Asha and I did not have a son during the first eleven years of our marriage must have been an additional source of disappointment to him, because most eastern cultures lay great store by the continuation of the male line.

It is a matter of some satisfaction to me, however, that our relations remained proper and there was never any overt unpleasantness. I was very careful to maintain all the respect that he deserved as head of the family. I wrote cordial letters to him regularly, to which he always replied, and we would spend a fortnight or so with him every winter. On my first meeting him during such visits I would touch his feet, and do the same when we took leave of him at the end of the stay. During our visits I would accompany him to the Mahalakshmi races as I had done in earlier days as a boy. He had suffered from diabetes for many years, but because of a curious needle-phobia he absolutely refused to take insulin injections. As a result, he developed a serious and extremely painful carbuncle on his neck in the late fifties. Although he kept it from us, we came to know about the ailment from his jeweller, Shanti Seth, and were deeply worried. When we saw him next we were shocked to see how thin he had grown. In fact, he was unable to walk and would go to the lift in a wheelchair when he emerged for his obligatory evening drive to the stables to see his beloved horses. He was then sixty-six.

On our way to England that year we spent five days with him in Bombay. Though unwell he was cheerful, and gave us detailed instructions on purchasing for him a pair of Manchester terriers, of which he had read recently and which were apparently bred in Scotland. He also wanted a pair of 8 × 30B binoculars for the races. Our flight left late in the evening, and I can distinctly recall taking leave of him on that occasion. I bent and touched his feet; I can still feel the texture of his smooth skin and hear his semi-jocular farewell, 'Be good, and look after Asha.'

We were in London for ten days, and although we did our usual shopping at Harrods and saw some plays, I was strangely uneasy. With considerable difficulty we were able to procure the pair of dogs that my father wanted, and I left it to my Secretary Dewan Iqbal Nath in London to bring them back to India while we broke journey for a few days in Italy. We had planned to visit Naples, and flew to Rome on 24 April, where we intended to spend two days before proceeding south. We were in a small hotel planning to take a bus tour of Rome, to my knowledge completely incognito, when to my surprise the phone rang. Our Ambassador in Rome, S. N. Haksar, was on the line saying that he was coming over immediately to see me. I realized that something must be wrong, and without telling Asha—who was unpacking—I went down and met him in the foyer.

'I have bad news for you,' he said, after we had shaken hands. 'Your father is dead.' The words hung in the air; he had been such a powerful presence all my life that it took me a while to realize that he was no longer alive. I broke the news to Asha, who was very attached to him and burst into tears. The Ambassador had most thoughtfully booked two seats to Delhi on an Air-India flight leaving Rome late that evening, so we hurriedly packed and rushed to the airport. That night on the plane I was unable to sleep and images crowded into my mind; my father, resplendent in uniform and medals, riding on a white horse at the head of his troops, inspecting the great Dussehra parade on the Batmaloo grounds in Srinagar; sitting in the Thricker fishing lodge after a successful day's fishing, sipping his whisky and laughing at the jokes made by his staff officers; cooking delicious recipes in the lovely lawns of Gulab Bhavan with my mother and a huge entourage; sitting next to me at the Bombay races shouting encouragement to his horse as the animals crossed the finishing line; beaming with pleasure when he won, angry and depressed when he lost. The old days in Kashmir came back to my mind, when we would spend a day out shooting duck or chakor. He was always a formidable, almost titanic figure in my consciousness. A distinguished reviewer of *Heir Apparent* perceptively remarked that I was craving for my father's love. I am sure he loved me, but yet I could wish that he had done so as I love my own sons, without inhibition or restraint.

We reached Bombay early on the 28th morning. As we emerged from the plane we noticed two dogs being off-loaded from the baggage compartment. By a strange coincidence, the Manchester terriers were on that very plane, but my father would never see them. We were met by his staff, who told us that he had suddenly taken ill on the 27th morning. A prolonged bout of coughing had developed into a heart attack. The doctor who was looking after him arrived almost at once and gave him injections, but they were of no avail. His last words were, 'Doctor, I am going.' Mercifully, his suffering was not prolonged, and an hour or so later he was dead.

In accordance with the instructions in his will, my father was cremated with Arya Samaj rites, and this had been done by the time we arrived. In a way I was glad not to have seen him dead. I went to the cremation ground and collected the bones, which were consigned to the sea in Apollo Bunder. Again, in accordance with the will, his ashes were placed in a silver urn. He had written: 'My *Asthis* should be thrown into the sea, but my ashes should be placed in a silver urn and taken up in a specially chartered plane and scattered over Jammu city, the place of my birth.' On 28 April we flew with the urn in a chartered plane to Delhi where, to my surprise, Krishna Menon, who was then Defence Minister, met me at the airport. After spending the night in Delhi, Asha and I flew on to Jammu the next day in the chartered DC-3.

Before landing we circled over the city thrice and I scattered the ashes out of a window over the soil beneath, the soil from which man springs and to which at last he returns. All of Jammu had gathered for the last homage—endless crowds, crying and wailing for a man who had left them a dozen years earlier as their Maharaja. And I carried the urn in an open car through the milling crowds. Thirty years ago on the same route my father and mother had driven in an open carriage through cheering throngs in celebration of my birth. Twelve years ago my parents and I were taken out in a similar procession after my return from America. This time I was alone; my mother had donned the white clothes of the traditional Hindu widow and lay semiconscious in Amar Mahal surrounded by wailing women, and all that was left of my father's body was a handful of ashes in the urn that I gently cradled in my hands. How much, in the end, does a man weigh!

Despite a mention in my father's will that no other religious ceremonies should be performed, I considered it my duty to go through the traditional thirteen-day period as laid down in the *shastras*, which involved a variety of prayers, sleeping on the floor, eating vegetarian food, once a day only, and not shaving. My mother, despite their estrangement and virtual separation for over a decade, took his passing very badly. She insisted upon observing the strict regulations for a traditional Hindu widow for a full year, as a result of which her health was severely affected.

On the conclusion of the thirteenth-day ceremony I wrote to Jawaharlal reminding him of the issue of the notification by the Home Ministry consequent to my father's passing away, and also sent a copy to Lal Bahadur Shastri, who was then Home Minister. This notification records the formal recognition by the Government of India of a successor when any of the former rulers died, a legal and constitutional requirement flowing from the accession of the princely States to the Indian Union. I had met Lal Bahadur on several occasions over the preceding years, but it was at this juncture that I came into closer contact with him. Diminutive, quiet and unassuming, he had a clear and shrewd mind, and an unusual capacity for sustained hard work. He was evidently close to the Prime Minister, and was his principal aide in several fields of party and national politics.

At Jawaharlal's suggestion I visited Delhi in June, and had discussions with Lal Bahadur. Clearly, the Prime Minister was not entirely easy with the whole question of princely titles, and thought that this would be a good occasion to start making some modifications. The State government was also allergic to the title of Maharaja being conferred on me, because of the Constituent Assembly's 'abolition' of the monarchy several years ago. In fact, my recognition as successor to my father was a constitutional necessity, but there was some discussion regarding the modalities to be adopted. Ultimately, at Jawaharlal's suggestion, conveyed to me through Lal Bahadur Shastri, I agreed not to use the title as long as I was *Sadar-i-Riyasat*. I wrote to him on 7 July 1961:

My dear Shastriji,

Although I have succeeded to the title of my late father His Highness Maharaja Hari Singh, I do not wish to use these titles so long as I hold the office of the Sadar-i-Riyasat of Jammu and Kashmir, and during this period I should like to be addressed as Shri Karan Singh.

With kind regards,

Yours sincerely,
K.S.

On the 10th the Government of India issued an official notification, signed by the then Home Secretary B. N. Jha, which read as follows:

ORDER

In pursuance of clause 22 of Article 366 of the Constitution of India, the President is hereby pleased to recognize Yuvaraj Karan Singh as successor to his late Highness Maharaja Sir Hari Singh Indar Mahindar Bahadur with effect from 26th April 1961.

This was accompanied by a press note which mentioned that the Government of India had been intimated that as long as I was *Sadar-i-Riyasat* I did not wish to use the titles of 'His Highness' and 'Maharaja' and would like to be addressed as 'Shri Karan Singh'. The note went on to say that 'the Government of India have noted with appreciation this request of Shri Karan Singh and will give effect to it'. As far as the privy purse was concerned, I was given the ten lakhs that my father had been receiving, out of which one lakh was earmarked for my mother for her lifetime. In view of the enhanced purse, I decided to discontinue drawing the salary of Rs 5,500 per month as *Sadar-i-Riyasat*, a gesture that was 'gratefully accepted' by the Council of Ministers headed by Bakshi Ghulam Mohammad.

Jawaharlal visited Srinagar again soon after I got back, and was given a rousing reception all the way from the airport to the Chashmashahi Guest House, including a colourful boat procession up the Jhelum with vast crowds thronging the banks and the historic bridges. In the evening he addressed a huge public meeting at Lal Chowk where, in his eight-minute speech, he dealt with a broad range of problems including the growing collusion between President Ayub Khan of Pakistan and the USA. He said that India refused to be intimidated by threats of violence from across the Ceasefire Line, and ridiculed the demand of a plebiscite from a country that had not held a single free election itself since it came into existence in 1947.

By now my activities in the State had grown considerably. I had turned thirty, and although I was playing a fairly active role as Head of State I was beginning to feel somewhat cramped and confined. While party politics was obviously not possible, I became more active in various other spheres. As Chancellor of Jammu and Kashmir University I presided over meetings of the Syndicate, helped to plan buildings at the hundred and twenty acre orchard that I had donated to the University at Hazratbal, and presided over the annual Convocation. I also attended Red Cross meetings, presided over meetings of the Board which looked after the interests of ex-servicemen, and generally tried to make myself useful in good causes while keeping a sharp eye on the political situation.

Asha was also busy. She continued her education after

passing the Bhushan (Hindi) examination in 1956 by taking the matriculation and intermediate examinations in English, and another Hindi examination—Prabhakar—in 1960. She was Chairman of the State Social Welfare Board, and in that capacity was involved in several projects in Srinagar city as well as in Jammu and Leh. The President of the Central Board was the able but formidable Durgabai Deshmukh. Asha was also Convener of the State Branch of the All-India Women's Conference. Her main aide in this work was a Kashmiri pandit widow Bageshwari Dhar, a dedicated social worker who had earlier been active with Mridula Sarabhai in recovering abducted women from Pakistan soon after partition.

Apart from the activities of State-level organizations, Kashmir was the venue of numerous all-India conferences and conventions, and I was often asked to inaugurate or address them, which gave me valuable opportunities to interact with different kinds of people. Towards the end of 1961 a session of the All-India Oriental Conference was held in Srinagar under the presidentship of Dr V. Raghavan, which was attended by a large gathering of scholars from all over the country. I found this meeting particularly interesting, and in my inaugural speech pointed out the unique position of Kashmir in the Indian cultural tradition, particularly in the field of Sanskrit studies.

While all this was happening, disturbing news began filtering in from Ladakh. The Chinese had already set up check posts in areas which legally were part of the former State of Jammu and Kashmir, and hence of India. In my fortnightly letter to the President dated 1 November 1961 I mentioned that the Chinese had set up another position at 'Chipchap' within India's borders and added: 'It seems fairly clear that the Chinese have no intention whatsoever of vacating the territory they have occupied, and are in fact busy stabilizing and consolidating their position there.' The military action in Goa took place early in 1962, but the euphoria of ending this last vestige of colonialism on the subcontinent was not destined to last very long.

The campaign for elections in Jammu and Kashmir started soon thereafter. The National Conference was supported by the Congress, and Jawaharlal paid a flying visit to Jammu to campaign for their candidates. But a curious incident took place which caused considerable embarrassment to Bakshi and his party. The Praja Parishad discovered that the ballot boxes could be opened easily without disturbing any of the seals or locks upon them. Such boxes were brought to my residence and the defect demonstrated in my presence. A short tap at a critical place would cause the lid to spring open miraculously, and this was widely reported in the national press. I immediately wrote to Bakshi and also contacted the then Election Commissioner, K. V. K. Sundaram. He flew to Jammu the next day and had a series of meetings with Bakshi, opposition leaders and local election officials. Thereafter he announced certain measures designed to make the boxes relatively more secure, such as pasting strips of paper outside each box and sealing them in cloth. However, the incident brought the whole electoral process under a cloud, and when polling in the Jammu region was held on 24 February and the National Conference won 26 of the 29 seats, the margin obviously did not reflect the mood of the people. In the valley, 34 National Conference nominees were elected unopposed as the papers of all opposition candidates were rejected, generally on flimsy grounds. In the result, while Bakshi won as many as 70 of the 75 seats in the new Assembly, the whole affair considerably damaged the credibility of the ruling party in the State.

In those days members of Parliament from Jammu and Kashmir were not directly elected, but were nominated by the State Legislature, one of the less attractive aspects of the 'special position' of the State. Six undistinguished National Conference men were so nominated to represent the State in the new Lok Sabha. The State Assembly was dissolved and Bakshi was again unanimously elected leader of the National Conference parliamentary party. I called upon him to form a new government, and a nine-member Cabinet headed by him and including G. M. Sadiq, Mir Qasim and others was sworn in by me.

Meanwhile, the third general elections had been held for the Lok Sabha and Jawaharlal Nehru had once again been returned to power. Asha and I visited Delhi at the end of March, where we bade farewell to Dr Rajendra Prasad who was relinquishing office after ten years as the President. His health had begun to fail, and often during Governors' Conferences he would suffer grievously from asthmatic attacks and had to constantly use an anti-allergic spray for his throat. A man of unfailing courtesy and genuine simplicity, he conducted himself with a kind of rustic dignity that was impressive. However, it was clear that he and the Prime Minister were steadily drifting apart. It was well known that Jawaharlal had not favoured his getting a second term as the President, but—true democrat as he was—had bowed to overwhelming opinion within his party. Dr Rajendra Prasad was in fact much too conservative for Jawaharlal's liking, but he never treated the President with less than scrupulous respect in public and in person.

While in Delhi we had occasion to meet two Nobel Laureates; Dr C. V. Raman, who delivered a fascinating lecture on 'The Gateways of Human Knowledge' at the Azad Bhavan, and Pearl S. Buck, who addressed an overflowing audience at the India International Centre auditorium, and later dined with us at Mansarovar along with Gertrude Emerson Sen. The only other woman of that stature that I had met, who was in some ways even more remarkable, was the legendary Helen Keller. Born blind and deaf, Keller was able to develop into an erudite and enlightened woman through a teacher who was truly a miracle worker. She had transited Jammu some years earlier, and I had met her at the airport. In Delhi on this occasion we also met the vivacious Jacqueline Kennedy at a dinner party given by Jawaharlal. Indira Gandhi had rung us up in Jammu asking Asha and me to come for the dinner, which we gladly accepted.

In May the scholar-statesman Dr S. Radhakrishnan was elected the President, a richly deserved honour in a country which has through the millennia reserved its deepest veneration for men of learning and wisdom. As the Sanskrit saying goes, a king is respected only in his own kingdom, but a scholar is respected wherever he goes. Appropriately enough, another gracious and cultured scholar, Dr Zakir Husain, was elected the Vice-President. Thus at that juncture the three highest elective offices in the land were held by men of outstanding ability and intellectual stature of which the nation could truly be proud.

Chapter 18

Later that summer Jawaharlal Nehru came to Srinagar again for a fortnight's stay, accompanied by Indira Gandhi. He had not been very well, and needed rest and a change. We left him alone as much as possible, though I did have one long meeting with him in which I discussed the political situation in the State. I had by then completed my doctoral thesis for Delhi University on the political thought of Sri Aurobindo (1893–1910), which was later published by Allen and Unwin in London under the title *Prophet of Indian Nationalism.* I gave the typescript to the Prime Minister with the request that he write a foreword. This he did, ending with the para, 'In commending this study of Sri Aurobindo's political thought, I would like to point out that the young man who has written it, is a scion of a princely house in India. At present, he is the elected Head or Governor of the State of Jammu and Kashmir. It is significant to note that he should have devoted himself to his studies and, more particularly, to this analysis of the political thought of Sri Aurobindo during a crucial period of our history, instead of occupying himself with the normal activities of Indian princes, that is, shikar, polo and high society.'

In fact, this was my third book. *Varied Rhythms*, a collection of poems and essays, had been brought out in 1960 by Asia Publishing House, and another book *Shadow and Sunlight*, a collection of Dogra-Pahari folk songs with musical annotation and English translations, was published in 1961. After my doctorate I began to be addressed as 'Dr' Karan Singh, a title I really much preferred to the feudal one because I had earned and not merely inherited it. Also that summer, I received a telegram from Dr Radhakrishnan appointing me to the honorary rank of Major General in the Indian army. I have often noticed that giving up something is far more satisfactory than hankering after it. I gave up one title and, within three months, had acquired two.

Talking of titles, at about this time I had become interested in the work of Albert Schweitzer, the remarkable doctor–

musician–philosopher who went to Equatorial Africa at the age of fifty and set up his celebrated hospital at Lambarene. *From the Edge of the Primeval Forest* made a deep impact on me, and I also read some of his other books. Although his treatment of Hinduism was not satisfactory, his philosophy of life-affirmation was impressive. A series of articles on him appeared in a Sunday newspaper, one of which concluded with the suggestion that he should receive some recognition from the Government of India. I wrote to Dr Radhakrishnan suggesting that 'it might be worthwhile to consider the institution of a special award for non-Indians, to be called perhaps "Bharat-Mitra", which could be awarded by the Government of India to outstanding personalities outside the country.' While the suggestion was not immediately accepted, it was fulfilled some years later with the institution of the Nehru Peace Prize. But that was after Jawaharlal had passed away, and in 1962 he was such a powerful presence that it was difficult to conceive of an India without him.

My encounter with Schweitzer was a stimulating one, although Bertrand Russel and Aldous Huxley remained my favourite Western authors. I wrote to Huxley enquiring about mescalin, the dramatic effects of which he had described so graphically in his small but influential book *The Doors of Perception*. I also said how much I had enjoyed his last novel *Island*. He wrote back:

> Thank you for your kind letter. 'Island' is a kind of pragmatic dream—a fantasy with detailed and (conceivably) practical instructions for making the imagined and desirable harmonization of European and Indian insights a fact. But alas, in spite of these pragmatic aspects, the book still remains a dream—far removed (as I sadly make clear in the final paragraphs of the story) from our present reality. And yet, if we weren't all so busy trying to do something else, we *could*, I believe, make this world a fit place for fully human beings to live in.

Alas, the ideal world of Huxley was in fact nowhere near attainment. Only fifteen years after the traumatic partition of the subcontinent, which resulted in tens of thousands of innocent people being killed and led to one of the largest mass migrations in human history, we were once again on the threshold of a violent clash, this time with our formidable eastern neighbour, China. After consolidating their position in Tibet, the Chinese had built up a strong military presence on the Tibetan plateau, from which they could dominate the incredibly difficult and tortuous mountain passes and valleys below. With a certain historical ambiguity built into our northern borders, both in the western (Ladakh) and the eastern (NEFA) sectors, which were a spillover from the imperial British legacy, there was a potential and permanent source of discord, of which I do not think the policy-makers in New Delhi were adequately aware.

It would have been in India's national interest to get the whole border with China unambiguously demarcated. However, despite all the euphoria about Indo-Chinese friendship, symbolized by the famous slogan 'Hindi–Chini Bhai Bhai', and the fact that Jawaharlal did all he could to get the People's Republic its rightful place in the United Nations, the border issue was never clinched. For several years there were minor incidents in both sectors. Chinese posts were set up in what we claimed to be part of Ladakh, and there were some probing moves by both sides in the NEFA area. In July 1962 there was an incident in the Galwan valley. Although the Chinese forces did not enter into a major conflict with our troops, it was clear that the absence of a mutually recognized border between the world's two most populous nations was going to create enormous problems. Simultaneously, there were reports of troop build-ups in Pakistan and Pakistan-occupied Kashmir, and incidents on the Ceasefire Line became increasingly frequent. There was also a flare-up in the eastern sector, and it was obvious that the Chinese were determined to reactivize the entire border.

The Indian army was one of the finest fighting forces in the world. It had proved its mettle in two World Wars, and after Independence it has been considerably strengthened. I had been on terms of close friendship with all the Army Chiefs since the withdrawal of the British—K. M. Cariappa, Maharaj Rajendrasinhji, S. M. Shrinagesh, K. S. Thimayya and the then Chief of Army Staff, P. N. Thapar. There had, however, been a qualitative and unfortunate change in the tradition of armed forces administration with the appointment of Krishna Menon as Defence Minister in 1956.

While his intelligence and ability were undoubted, he was responsible for introducing an element of politicization into the army that within a few years had disastrous repercussions. Strong-willed, arrogant and suspicious to the point of paranoia, Krishna Menon fully exploited his closeness to Jawaharlal by pushing his own favourites in the Defence Ministry and, worse, in the services. When he took over the Ministry, General Thimayya was Chief of Army Staff. A man of candour, with an impressive personality, he clashed almost immediately with Menon and this ultimately led to Thimayya's abortive resignation in 1959. Jawaharlal backed Menon to the hilt, Thimayya was humiliated and forced to withdraw his resignation, and when his term expired he was succeeded by General P. N. Thapar, an amiable but much weaker man than his predecessor.

Krishna Menon, however, was not content to work through Thapar. He pushed his favourite General B. M. Kaul, who, as a Brigadier, had functioned as a liaison between Srinagar and Delhi during the political crisis of 1953 which resulted in Sheikh Abdullah's dismissal and imprisonment. 'Bijji' Kaul was a highly intelligent and emotional person who was close not only to Menon but also to Jawaharlal. He was appointed Chief of the General Staff and, in fact, began running the army. It was this unfortunate situation—an ageing Prime Minister, an arrogant and power-oriented Defence Minister, a weak Army Chief, a crafty and over-ambitious Chief of the General Staff—that was largely responsible for the debacle that followed. By this time all of Jawaharlal's senior colleagues and contemporaries in the government and Congress Party had disappeared. There are obvious advantages of being on a lofty pinnacle of power, but the disadvantages and dangers are also equally considerable. One of them is that the unquestioned leader does not have people around him who talk frankly and fearlessly. He is told what those close to him think he would like to hear, as a result of which he gets increasingly isolated from reality and starts living in a make-believe world of his own creation. When reality intervenes, it is invariably a personal and national trauma.

As I have remarked earlier, the Government of India was not fully aware of the complexities of our northern frontiers, in

Ladakh or in NEFA. The British had for two centuries controlled the whole region with a view to safeguarding their imperial interests. With their withdrawal, India found itself embroiled in partition, and did not really have the time to develop its strategic or diplomatic expertise *vis-à-vis* China. This explains both the diplomatic debacle in Tibet and the military debacle of 1962. We clearly fell between two stools, being unable on the one hand to strike a diplomatic settlement with China and on the other to organize sufficient military strength to enforce our claims. There was insufficient realization of just what a Himalayan expedition involved, and there was no Zorawar Singh at Army Headquarters with the military genius to overcome India's intrinsic weakness in that area.

On the morning of 20 October Chinese troops launched strong attacks both in Ladakh and in NEFA. At the Governors' Conference on the 23rd, Jawaharlal gave a detailed account of the situation. He conceded that we had suffered a severe setback, that many of our posts had been overwhelmed and that the Chinese were advancing towards Tawang in the west and Dungli in Ladakh. The topography was, in fact, entirely favourable to them, and our forces were given impossible tasks without the necessary logistic backing. The 'forward strategy' that Kaul and Menon between them had sold to Jawaharlal collapsed overnight. Not only did we not move forward, we were in fact pushed back all along the border and suffered heavy casualties. It was the darkest hour for our armed forces, under-equipped and hopelessly outnumbered. The nation was stunned. Having been led to believe that its borders were secure, it suddenly realized that in fact they were utterly vulnerable, and that the once impregnable Himalayas were no longer the impenetrable barrier and sentinel that they had been from the dawn of civilization.

Immediately after the Governors' Conference I flew back to Srinagar, because 28 October had been fixed as the last date for filing nomination papers for the five-yearly election to the office of *Sadar-i-Riyasat*. The only nomination paper received was mine; it was proposed by Bakshi Ghulam Mohammad and seconded by Girdharilal Dogra. On the 29th morning, the Speaker announced that I had been unanimously re-elected *Sadar-i-Riyasat* for the third time, and I took the oath of office on the 31st. Despite the Chinese attack, morale in the State was high. Bakshi, always good at crisis management, made fighting speeches in the valley and in Jammu. I flew down to Jammu immediately after the swearing-in, and found the people full of enthusiasm for national defence and unity, a phenomenon that was in evidence throughout the length and breadth of the country. Collections for the National Defence Fund were in progress.

After two days in Jammu I went back to Delhi. The Chinese invasion had transformed the political landscape. While Jawaharlal himself had ridden the storm, his prestige at home and abroad had been seriously undermined. As it became clear that we were being decisively worsted in the border war, that our army was ill-equipped and inadequately clothed to fight effectively in the bitter cold of that inhospitable terrain, that Krishna Menon and General Kaul between them had led us into a military and political disaster, the public mood began to change. While there was tremendous patriotic fervour, and a brave resolution was adopted in Parliament, there was growing anger at all levels including the Congress Parliamentary Party.

The demand grew that Krishna Menon must go. Jawaharlal, loyal to a fault and always unduly partial towards Menon, resisted for a while. On 31 October he announced that he himself was taking over the Defence portfolio but that Krishna Menon would stay on in the Cabinet as Minister for Defence Production. Clearly this was not enough. In any other country the Prime Minister himself would have had to go, but such was Jawaharlal's prestige at home and abroad, his unique record of public service, and his transparent sincerity that the demand was not raised. The only parallel I can think of was what happened in Egypt with President Nasser after the crushing defeat in the six-day war. However, Jawaharlal's attempts to save Menon were simply not feasible at that stage. On 7 November there was a stormy meeting of the Congress Parliamentary Party, where it became clear that, if he continued to shield Menon, the party would even be prepared to consider a change in the leadership itself. Some years later the witty and ebullient Mahavir Tyagi, who was one of the main leaders of the anti-Krishna Menon faction, described to me the events of that meeting in the Central Hall. He said that Jawaharlal first praised Krishna Menon for several minutes for his ability and the services he had rendered to the nation and that at each mention the small, pro-Menon lobby would pound triumphantly on their tables. Then at last the Prime Minister said that Menon had tendered his resignation and that he had very reluctantly accepted it. No sooner were these words uttered than most of the members thunderously thumped their tables, and this went on and on until Jawaharlal finally got up and the meeting came to an end.

Whatever other intentions the Chinese had, they did succeed in gravely damaging the prestige and power of Jawaharlal. Overnight his whole situation both within the country and *vis-à-vis* the outer world was transformed. After the ouster of Krishna Menon, General P. N. Thapar resigned and was replaced by General J. N. Chaudhuri. Bijji Kaul also resigned and spent the rest of his life a broken man. Jawaharlal brought in as Defence Minister the dynamic and successful Chief Minister of Maharashtra, Y. B. Chavan, who arrived in Delhi the very day China announced its unilateral ceasefire. The USA suddenly became friendly, and President Kennedy wrote to Jawaharlal saying, 'Our sympathy in the situation is wholeheartedly with you. You have displayed an impressive degree of patience and forbearance in dealing with the Chinese. You have put into practice what all great religious leaders have urged, and so few of their followers have been able to do.' This was clearly a tactful overstatement. In fact, India had been decisively trounced in the first border war with its eastern neighbour. Having personally guided our foreign policy since Independence, Jawaharlal could not and did not seek to evade responsibility for the debacle. The Soviet stand was clearly ambiguous, as at that time the rift between the two communist giants had not fully surfaced.

For almost a month there was a flurry of conflicting reports in the press, but it had soon become clear that we were losing ground both in the eastern and the western sectors, and at one point it seemed that the Chinese might well pour down from the Himalayas into the thickly inhabited Brahmaputra valley and the Indo-Gangetic plain itself. Jawaharlal in anguish admitted that he had been 'getting out of touch with reality in the modern world and living in an artificial atmosphere of our own creation'. American and British arms started arriving in Delhi, along with some military experts. Given his well-known antipathy to any Western military presence in India, this was a measure of Jawaharlal's helplessness in the face of a conflict that caught him by surprise.

The big question mark during the weeks of the war was whether Pakistan would join the fray. Luckily it did not, but, once again, along with military aid the Western powers started exerting strong political pressure upon India to come to a settlement with Pakistan over Jammu and Kashmir. Averell Harriman visited Delhi on 22 November, one day after the Chinese had announced a unilateral ceasefire, along with the British Minister, Duncan Sandys. Both put a good deal of diplomatic pressure on Jawaharlal to agree to some settlement on Kashmir, and I can recall that all sorts of plans were in the air including even the partition of the valley. Despite his weakened position, however, the Prime Minister refused to budge on this, but he was obliged to agree to reopening negotiations with Pakistan. Over the next eight months several rounds of talks were held between the Indian delegation led by Sardar Swaran Singh, surely one of the world's greatest experts at diplomatic stonewalling, and the Pakistani delegation led by the brilliant but mercurial Zulfikar Ali Bhutto. I cannot think of two personalities so utterly different, which of course symbolized the fact that the views of the two countries were poles apart. It is a measure of their total lack of realism that Pakistan seriously put forward a proposal whereby India would be left only with the Kathua district in the Jammu province, that is, with about 3000 out of the total of roughly 85,000 square miles of the State of Jammu and Kashmir. India was prepared to make certain adjustments in the Ceasefire Line if a permanent settlement could be reached, but for Pakistan the Kashmir valley had a fatal attraction that was ultimately to destroy that nation itself. No agreement was reached either by direct negotiations or, later, at the Security Council when the Soviet veto once again came to our rescue.

Back in Jammu, we were all in a state of suspense for weeks, half expecting an attack across the border from Pakistan. A 21-member State Citizens' Council was set up with me as Patron and Bakshi as President, and various steps were taken to mobilize resources for the national defence fund and to raise a Home Guard of twelve thousand people. Asha and I visited military hospitals in Jammu and Udhampur, where the wounded from the Ladakh front were steadily coming in. I had got my military uniform prepared after becoming an honorary Major-General, and appeared in it for the first time during the war period, while Asha's Ladies Club in Jammu and Zenana Club in Srinagar had started a widespread campaign to collect clothes and foodstuff for the armed forces in hospitals and at the front.

Looking back at the entire affair, three things seem to be clear. First, the Government of India with all its expertise and experience was not really adequately aware of the complexities of the Himalayas. Militarily we were entirely obsessed with Pakistan, and our strategic thinking was geared towards the threat from there. Having lost an opportunity in Tibet, we fell into a perilous situation both diplomatically and militarily. Jawaharlal, after all he had done to befriend the People's Republic of China, virtually introducing Chou En-lai to the international community and consistently working for China's seat at the United Nations, never realized that when it came to what they considered to be their vital interests, the Chinese would be ruthless and unbending. His own idealism and essentially liberal-democratic upbringing, as well as a somewhat idealized image of communism, prevented him from seeing, until it was too late, the real face of a totalitarian regime not bound by the checks and balances of a democratic polity.

Secondly, Jawaharlal was badly advised or, rather, chose to take advice from people who did not have the correct assessment. On the political front, Krishna Menon was the main architect of disaster, as he misled the ageing Prime Minister into a totally wrong handling of the issue. His constant statements that the Chinese would be 'thrown out' or 'thrown back' if they crossed what we considered was the international border were highly irresponsible, because on the ground we were militarily in a far weaker position than the Chinese—topographically, manpower-wise and as far as equipment and training were concerned. Sabre-rattling is a disastrous exercise if the scabbard has only a dagger instead of a sword. And, as I had seen once many years earlier with my father, and was to see once again many years later under different circumstances, too much power concentrated in a single individual is ultimately counter-productive.

Thirdly, the brief interlude in which Krishna Menon politicized the army compounded the disaster. The blatant manner in which General Kaul was boosted and factionalism introduced at all levels in the armed forces could not but have tragic repercussions. In a way it was good that the process was exposed; another few years and the rot would have seeped down to the roots, permanently crippling our magnificent fighting forces. As it turned out, this rude shock had one incidental advantage. It shook us out of our complacency and gave us a much more realistic appreciation of the military and strategic responsibilities that a power of India's size and location inevitably had to shoulder.

In many different ways the 1962 India–China war marked a turning point in Asian and, indeed, world history. In India it marked the beginning of the end of Jawaharlal Nehru's long and outstanding career. He never got over the political and personal shock suffered during those traumatic weeks. Physically he seemed to shrink in some way, and age rapidly. Although his voice and demeanour retained its fibre, the event had dealt him a blow from which he would never recover. For me, who had always been his ardent admirer, it was excruciating to see this happening. Better, I thought, to go dramatically like Gandhiji rather than to have to face this humiliation. But the ways of destiny are inscrutable, and the subtle threads of *karma*, invisible yet stronger than the finest steel, make no distinction between prime ministers and other mortals.

Chapter 19

I was ex-officio Chancellor of Jammu and Kashmir University, but in 1960 I was also elected Pro-Chancellor of Banaras Hindu University, a great centre of learning founded by Pandit Madan Mohan Malaviya some decades earlier. The Chancellor at the time was the Maharaja of Mysore, but he seldom attended convocations and so I had to preside over these annual functions. It was always a pleasure to visit Varanasi, where the temporal and the eternal intersect in a most dramatic fashion. The first year we were guests of the well-known cricket enthusiast 'Vizzy', the Maharajkumar of Vizianagaram, who lived in a palatial mansion near the University, and later of Maharaja Vibhuti Narayan Singh of Banaras, a man of learning and dignity.

The University itself is a splendid complex of buildings, probably the best planned campus in India. In 1963 I was elected Chancellor of the University, and a visit to Varanasi for the Convocation became an annual affair. The Vice-Chancellor then was N. H. Bhagwati, a distinguished jurist retired from the Supreme Court. The Convocation was conducted in Sanskrit, and the whole function had a touch of saffron that made it unusual and attractive. My mother came with me once, and also visited the sacred spots near Allahabad. Earlier she had gone to Haridwar and stayed for a month at our beautiful *haveli* on the banks of the Ganga. She also went on a pilgrimage to Badrinath, thus fulfilling a life-long ambition. Always fond of colourful clothes, she had retained a village girl's rather naive love of heavy jewellery, but after my father's death she wore only white and turned increasingly towards religion.

Soon after becoming Chancellor of Banaras Hindu University I was invited by Aligarh Muslim University, where the degree of Doctor of Law (*Honoris Causa*) was conferred upon me. In the evening I addressed a meeting of the University Students' Union where I spoke on the various dimensions of the conflict with China—military, diplomatic, economic and ideological.

The speech was well received, and I was happy to be associated with this great University founded by Sir Saiyyed Ahmed. I felt that these two Central Universities at Varanasi and Aligarh could develop a much more cohesive and creative relationship with each other. I wrote to Dr Radhakrishnan about this and also mentioned it to Dr D. S. Kothari, the distinguished scientist-educationist who was then Chairman of the University Grants Commission.

The visits to Varanasi deepened my interest in the philosophical aspects of Hinduism. I was particularly attracted to the Upanishads, those marvellous spiritual dialogues that took place several thousand years ago. There is a popular misconception that the Upanishads are too difficult and abstruse for the general reader to follow. This is not true, particularly of the Upanishads in verse that combine the most beautiful poetry with brilliant insights into the nature of reality. The *Isha*, *Kena*, *Katha*, *Mundaka* and *Shwetashwatara* Upanishads, in particular, I found to be of great value as sources of spiritual insight as well as for the sheer aesthetic beauty of the language. On one occasion in Varanasi I ventured to lecture on the *Mundaka* Upanishad, with its famous lines:

> Having taken as a bow the great weapon
> of the secret teaching,
> One should fix to it the arrow sharpened
> by constant meditation.
> Drawing it with a mind filled with
> the Divine
> Penetrate, O good-looking youth, that
> Imperishable as the target.
> The Pranava (Aum) is the bow; the
> arrow is the self,
> Brahman is said to be the target.
> With concentration is It to be penetrated,
> One should become one with It as the
> arrow with the target.

The same text has the celebrated words *Satyameva Jayate*—Truth Alone Triumphs—which we adopted as our national motto after Independence.

While in Varanasi I also had occasion to meet outstanding scholars like Vasudev Sharan Agarwal, Pandit Gopinath

Kaviraj and Thakur Jaidev Singh, and gained a deeper appreciation of the immense power of intellectual and spiritual attainment that underlies the Hindu tradition. Some of the surface manifestations may be unpleasant, such as the half-burnt bodies floating in the Ganga, but to appreciate the strength of Hinduism one has to look below the surface and see the great current, like the Ganga itself, that has flowed unbroken from the dawn of history down from the mighty Himalayas to the boundless ocean.

My main spiritual inspiration remained Sri Krishnaprem and Sri Madhava Ashish. After that first visit in 1958 I went up to Mirtola regularly every year, and as our meetings became more frequent our correspondence also developed. Sri Krishnaprem seldom wrote in those days, but I have several letters from him which are among my cherished possessions, and I found his guidance on many matters invaluable. He laid considerable emphasis on dreams which he believed, along with Jung, carried important signals to us from our subconscious selves. If properly interpreted, these could give valuable indications on our inner situation, particularly our emotional problems. He laid stress on an integrated development of the head and the heart, the intellectual and the emotional, the male and the female aspects of the personality.

I happen to be a prolific dreamer myself, often getting through three or four full-length episodes in a single night. I developed the practice of writing down the more vivid of these dreams, which I would send to Mirtola. Sri Krishnaprem or Sri Madhava Ashish would return them to me with their comments, and thus I gradually embarked upon an experiment in introspection which turned out to be of considerable value in my inner growth. Many of my dreams obviously reflected the political situation, because important political figures and events figured prominently in them. But there were others with a deeper and more interesting aspect. Apart from clearly symbolic, archetypal situations such as swimming across stormy waters, climbing steep mountains, driving over bridges which span torrential rivers, visiting mysterious temples and so on, my dreams would often involve encounters with Hindu saints and deities. I had a particularly vivid dream in which Sri Aurobindo and the Mother initiated me into some form of higher consciousness as a result of which I went into a trance, and another in which Sri Ramana Maharshi beckoned to me to join him. On two occasions I have seen Krishna—once as a baby, and the second time as a youth—and have also encountered the goddess in various forms and disguises. This rich psychological material proves, if proof were needed, that waking consciousness is only a small fraction of total awareness. Above and below the conscious mind are vast dimensions, of which few of us have more than a vague inkling. Indeed, an important aspect of the spiritual path involves a deliberate effort to expand the borders of consciousness, and working over one's dreams is a valuable exercise in that direction.

Sri Krishnaprem laid special emphasis on the emotional realm, where problems had to be squarely faced and not brushed under the carpet, as it were, to fester and surface later as neurotic manifestations. He felt that while my intellectual development had proceeded rapidly, there had not been a corresponding emotional growth. He made it his business to correct this, and when an occasion arose of what he called 'a brush with the wings of Eros', his sympathetic but firm handling of the situation was of tremendous value. In one of his letters he said, 'A friend once wrote "love-proof" in paradise—what a disaster.' I think he was quite right, though perhaps he did not go far enough. To be 'love-proof' *anywhere* is a disaster—in fact *the* disaster. It is one of our troubles that we are so seldom able to integrate our love into what we conceive to be our spiritual life. We are too apt to think of the latter as an affair of *japa* or other such practices on a background of 'lofty' philosophy, and to forget that life without love can never be the fuller integrated life of the spirit to which nothing whatever that spirit has created can be alien. The spiritual life is essentially a *real* life concerned with *real* things, not with the substitutes of ritual and even most of what is called 'meditation'. First among these real things is love; when a man has it, he wants *nothing* else. It is the one-without-a-second. Many people will agree to this, but with the proviso that it is what they call 'divine' or 'spiritual' love, as if that power which has called forth the universe, 'the breath whose smile kindles the universe', could ever be undivine or that which is essentially spirit could become unspiritual!

Sri Krishnaprem's letters, his personal contact and his marvellous books, *The Yoga of the Bhagvadgita* and *The Yoga of the Kathopanishad*, remain a major element in my life. Whether in pain or in joy, in elation or in depression, in conflict or in repose, his fragrance lingers like a tangible presence and gives me strength. Along with Swami Vivekananda and Sri Aurobindo, he constitutes the trinity of remarkable men who have influenced me most in my spiritual quest, and he was the only one of them whom I had the privilege of meeting in the flesh.

Meanwhile the Indo-Pakistan talks in Karachi were followed by a visit to Peking by Z. A. Bhutto, during which a treaty was signed whereby Pakistan ceded to China about two thousand square miles of territory in the northern region of Jammu and Kashmir under its control, thus initiating a collaboration that was to have momentous implications which continue down to this day. Reeling as we were under the shock of the disastrous border war with China, we were not in a position to do very much except register the usual protest; but the National Conference Working Committee meeting in Srinagar in May passed a long resolution reaffirming the stand that Jammu and Kashmir was and would remain an integral part of India. Jawaharlal Nehru came up for a ten-day stay in June, most of which he spent in Pahalgam. While he retained his smile, and received an enthusiastic welcome on his arrival with the usual cheering crowds all the way from the airport to Chashmashahi Guest House, he had begun showing signs of age and the tremendous strain that he had been undergoing for the last few months. Indira Gandhi was with him, and I found it much easier conversing with her. She was always particularly interested in Kashmir, and would react constructively to my suggestions. She never liked Bakshi Ghulam Mohammad, and found herself ideologically closer to G. M. Sadiq and D. P. Dhar, who had a more leftist approach to economic problems.

She found our company interesting, and we would discuss a number of political and other matters informally. She often read some of the books in our Srinagar library. I have a note from her about one of her visits:

Dear Tiger,

My father would like you to come tomorrow morning. Would 10 o'clock suit you?

I enjoyed my tea with you and your very beautiful wife. The word beauty is used often and far too casually but she really has a quality for which there is no other word.

I am returning Priestley's book. It is interesting and the situation it describes is not, I think, outside the realm of possibility. May I borrow Spear's book on India and Hanley's book?

Greetings,

Sincerely
Indira

In fact with Jawaharlal's failing health, Indira Gandhi's influence in the political life of the country grew steadily. Not only was she the closest person to her father, she was also keenly interested in politics, having been literally born into public life when the founding father of the Nehru dynasty—Pandit Motilal—was a major figure in Congress politics, and having subsequently grown up under the tutelage of her father. There was a widespread misconception at the time that she was an incidental figure who derived all her influence from the fact that she was Jawaharlal Nehru's daughter. However, for those of us who came to know her well, it was clear that she had a deep and abiding interest in politics, and that she would one day play a prominent role after her father's departure. The Nehrus were always proud of their Kashmiri ancestry, and felt a special kinship with the beautiful valley and its people. Jawaharlal had dwelt on this with feeling and eloquence in his writings, and Indira Gandhi had inherited his partiality for Kashmir. From the time of the Chinese invasion I had begun meeting her more frequently and discussing the internal situation in the State. Unfortunately, over the years Bakshi had become authoritarian and his regime increasingly corrupt. I was not happy at this, and made no secret of my feelings to Indira Gandhi, who shared much of my apprehension. As it turned out, the whole situation developed in one of those unexpected ways which defy all logic and which, indeed, give politics its peculiar unpredictability.

In August 1963 Jawaharlal Nehru sponsored a major political move that came to be known, quite erroneously, as the

'Kamaraj Plan'. In fact, K. Kamaraj, though successful in his home State of Tamil Nadu, found himself totally at sea in the arena of national politics. Fluent neither in English nor in Hindi, his effectiveness was sharply reduced, and given the towering figure of Jawaharlal Nehru as Prime Minister, his scope for creative political initiative was minimal. Over the last few years Morarji Desai had worked himself up to a position where he was clearly number two in the Cabinet, both by virtue of his experience and political stature. With the Chinese debacle, and his own health beginning to falter, Jawaharlal evidently began seriously considering the question of who would succeed him. Morarji was ideologically and personally not in harmony with the Prime Minister, being too conservative and rigid in his economic and political thinking to suit Jawaharlal's more leftist stance. I have no doubt that though the Kamaraj Plan, whereby senior Cabinet Ministers and State Chief Ministers were to resign and 'work for the party', was presented as a move to revitalize the Indian National Congress by injecting new blood at senior levels, the main political goal was to oust Morarji Desai from a position whereby he would almost automatically step into the Prime Ministership if something were to happen to Jawaharlal. This was a prospect which was unacceptable not only to Jawaharlal himself but also to many other Cabinet Ministers—including Lal Bahadur Shastri—and to Indira Gandhi. I also found Morarji to be a rather forbidding figure, and did not have with him the close personal rapport that I had developed over the previous fifteen years with Jawaharlal and Indira Gandhi.

The Congress did not at that time exist in Jammu and Kashmir, though the ruling National Conference was closely allied with it, and we assumed that the plan would not apply to the State. Bakshi Ghulam Mohammad went to Delhi soon after the plan was announced and said at the AICC meeting that, though strictly speaking he was not a Congressman, he would also fall in line and submit his resignation. Even then we assumed that it was simply a grand gesture that would not be taken seriously. When the news came on the evening of 24 August that the resignation of six Central Ministers and six Chief Ministers, including that of Bakshi Ghulam Mohammad, had been accepted, we were all stunned. Bakshi told me later that he had pressed his resignation because he was the only Muslim involved, and as he had played an important role in the discussions on the Kamaraj Plan he felt that his resignation would strengthen the Prime Minister's hands. While this was difficult to accept at its face value, I quite frankly could not come up with a better explanation. That Jawaharlal would be happy to ease him out was understandable, but whether Bakshi acted out of a sense of loyal idealism, or whether he never really thought that he would be allowed to go, is something that is difficult to prove one way or the other.

After the initial surprise, two trends developed among politically active groups in the State. On the one hand, the ruling National Conference, under its General Secretary, Bakshi Abdul Rashid, launched a widespread campaign of processions, demonstrations and resolutions protesting against the resignation and demanding that Jawaharlal and Bakshi should reconsider the decision. Several deputations met me urging that I intervene on the plea that Bakshi's departure would have disastrous political repercussions. On the other hand, some opposition groups, including the pro-Pakistan Political Conference, took the opposite view and urged that the resignation should stand. Significantly they took out processions raising pro-India slogans, and some of their leaders sought to assure me that although they had once been pro-Pakistani they had now changed their views and were staunchly pro-Indian. The Praja Parishad in Jammu took the stand that it was not interested in personalities but hoped that all distinctions between Jammu and Kashmir and other States should be abolished.

The pro-Sadiq group in the National Conference was delighted with the resignation, but maintained a low profile as the parliamentary and organizational wings of the party were both firmly under the control of the two Bakshis. Bakshi did say to me that in his view G. M. Sadiq was the best person to succeed him, but either this was simply a facade or he was in fact not able to control his headstrong cousin Abdul Rashid whose strong-arm methods had over the years brought his regime into disrepute. The whole situation was a delicate one, and as *Sadar-i-Riyasat* I was once again put in a difficult position. As I wrote to the President in my fortnightly report of 2 September 1963:

> Broadly speaking the decision taken jointly by the Prime Minister and Bakshi Sahib, though it will remove from the Prime Ministership of the State a man with remarkable qualities of political leadership and personal dynamism, should be viewed positively both as a challenge and an opportunity. It is a challenge in that there are always elements at work in Kashmir to disrupt peaceful life and create disturbance with a view to furthering various political ends that militate against the national interest. These must be carefully watched and countered. On the other hand this dramatic move does provide an opportunity to bring about certain changes in the internal affairs of the State which can in the long run lead to considerable political consolidation. It is a delicate and tricky situation, and I am looking forward to the guidance of yourself and the Prime Minister in the adequate discharge of my constitutional responsibilities as Sadar-i-Riyasat.

The crisis dragged on for several weeks. The National Conference Working Committee and legislative party held several meetings, the air was thick with rumours, and the State administration came virtually to a standstill. Clearly the ruling clique was not prepared to give up political power and all that went with it, and Bakshi Rashid began calling the tune. In my view Bakshi Ghulam Mohammad had resigned in a genuine mood of idealism—even his bitter opponents would concede that he did have some rare human qualities—but on his return to Srinagar his cohorts virtually pressurized him into either withdrawing his resignation, which he resolutely refused to do, or electing some pliable figure-head as leader who would in effect carry out the dictates of Abdul Rashid and his henchmen. In this they were joined by his second wife, Khurshed, whom Bakshi had quietly married recently. Rashid himself was far too unpopular with the general public and with Delhi to risk seeking election himself, but clearly he was not prepared to tolerate Sadiq.

Jawaharlal announced at a press conference early in October that Bakshi favoured the election of G. M. Sadiq as his successor, and this was also what he had told me. However, Bakshi was simply not willing or able to carry the party with him. He flew to Delhi on 9 October to attend a meeting of the

Congress Working Committee, and also had two meetings with the Prime Minister in which he indicated his inability to get Sadiq elected. On the 10th he returned to Srinagar, and a meeting of the National Conference Parliamentary Party was held that afternoon. Of the 101 members in both the houses, 86 were present because Sadiq and his colleagues did not attend. Bakshi proposed the name of Khwaja Shamsuddin, Revenue Minister in the outgoing Cabinet, as leader, and this was seconded by Bakshi Rashid. Shamsuddin was elected unanimously, and in his acceptance speech he said that his greatest ambition was to follow in the footsteps of Bakshi Ghulam Mohammad, and added that the day Bakshi agreed to return as leader of the party would be the happiest day in his life.

The report of the Parliamentary Party meeting was conveyed to me in writing by the Parliamentary Party Secretary, Sardar Kulbir Singh, who attached a list of the 86 members who had voted. Thereupon I wrote to Shamsuddin calling upon him to form a new government, and on the 12th morning administered the oath of office to a new Cabinet headed by him and consisting of six other Cabinet Ministers and four Ministers of State. Thus, under curious circumstances, ended a full decade of Bakshi Ghulam Mohammad's stewardship of the State, beginning on 9 August 1953 when, under even more dramatic conditions, I had administered the oath of office to him after the dismissal and arrest of Sheikh Abdullah. During his ten-year rule there is no doubt that the pace of economic development in the State was speeded up, Jammu and Ladakh given a somewhat fairer share of developmental activities, and several significant measures taken to strengthen the constitutional, legal, financial and administrative links between the State and the rest of India. All this goes to Bakshi's credit, and his patriotism was never in doubt. He also had some endearing human qualities, including excellent personal relations with Muslims as well as Hindus, and a generous if emotional temperament.

On the debit side, Bakshi's regime became increasingly authoritarian and corrupt as the years went by. His relations were particularly active in making money, and brought him a good deal of public resentment. When relatives and favourites begin to misbehave it is only a matter of time for the main person to get associated in the public mind with their misdeeds. Unfortunately, Bakshi had alienated the more enlightened sections of the intelligentsia, and depended largely on lumpen elements who could carry things off for a while but were ultimately doomed to collapse. Let it not be forgotten, however, that Bakshi rendered signal service to the nation at a very crucial juncture. Had he not stepped into the breach after the dismissal of Sheikh Abdullah and consolidated the nationalist forces in the State, we would have been faced with a much more difficult situation. The process of integration was continued by him right down to the end.

A few days before he relinquished office he sent me a memorandum dated 3 October 1963 in which he informed me that he had made two important political announcements during a debate in the Legislative Council, both of which we had been favouring for a long time. He wrote:

> As this was to be the last session of the Legislative Council to be attended by me as Prime Minister of the State I made an announcement of the following policy directives for implementation by the successor Government:
>
> (a) that the designation of the Head of the State (Sadar-i-Riyasat) and that of the Head of the Government (Prime Minister) should be changed to 'Governor' and the 'Chief Minister' respectively, as in other States of the country; and
>
> (b) that there should be direct elections to the Parliament instead of the present system of nominations—of course after the present Emergency is over.

The changes he proposed would remedy two glaring anomalies; it was patently absurd to have two Prime Ministers in the same country, and, as for the office of *Sadar-i-Riyasat*, nothing would please me more than being the first and last person to hold that office. Despite all the obloquy heaped on Bakshi later, an occupational hazard in the grim and pitiless political processes, his very real national contribution should always be remembered with gratitude.

His successor Shamsuddin was a much less substantial figure. He found himself suddenly catapulted into the Prime Ministership of the State, an office for which he was politically and psychologically quite unprepared. Despite some interesting though outlandish ideas about moral correction camps for corrupt officials, he never really found his feet, and within four short months he was swept away in the aftermath of the next crisis that confronted us. Whatever else may be said about being Head of State, there was never a dull moment.

Chapter 20

Soon after the Shamsuddin Ministry took office, the tempo of political activity in the State, inhibited during the Bakshi decade, began to intensify. The Praja Parishad, led by Pandit Prem Nath Dogra, was always closely linked with the late Dr Shyama Prasad Mukherji and the Bharatiya Jana Sangh, and it decided to formally merge with that party. Thus, for the first time an all-India party would be represented in the State Assembly. The State unit of the Praja Socialist Party was led by Balraj Puri, who met me to discuss the political situation, as did Shafi Qureshi, who led a new party called the Jammu and Kashmir National Congress. Both complained that Opposition parties in the State were not being given facilities to function in a democratic manner.

The Governors' Conference was held as usual in Rashtrapati Bhavan that winter, for the first time under the Presidentship of Dr S. Radhakrishnan. Over the years the old Governors had gradually retired, their term being five years, and as seating was on the basis of seniority I found that I had reached the head of the table and took my place immediately after the President, Vice-President Zakir Husain and the Prime Minister. It was faintly embarrassing to find myself at thirty-two seated above such venerable veterans as V. V. Giri, Governor of Kerala, H. V. Pataskar (Madhya Pradesh), Bishnuram Mehdi (Tamil Nadu), Biswanath Das (Uttar Pradesh), Dr Sampurnanand (Rajasthan), Ananthasayanam Ayyangar (Bihar) and Pattom Thanu Pillai (Punjab). But, apart from sitting next to Jawaharlal, there was also the pleasant fact that Padmaja Naidu, second-term Governor of West Bengal, was to my right. She was always great fun, and after the two of us had made our presentations we would exchange poems and notes.

These conferences had by then become rather boring. Jawaharlal, after listening to Padmaja and me, would often doze off. He was evidently tired and found the situation far from interesting. On one occasion, with her irrepressible sense of humour, Padmaja said to him, 'Jawahar, you should have these conferences more frequently.' 'What do you mean', he shouted, 'isn't once a year enough?' 'It is,' she replied, 'but you get such a good rest during them that it would do you a lot of good to have them twice a year.' She had inherited her irreverent sense of humour from her distinguished mother, Sarojini Naidu, who in the early days of the freedom movement once described Motilal Nehru, Jawaharlal and Gandhiji as 'the father, the son and the holy ghost'.

I had started writing poetry in the early sixties, and I wrote a satirical poem during the conference that was published later in a collection of my poetry *Welcome the Moonrise.* It was called *The Conference*:

In trudge the white-maned elders,
some with no manes at all—
flamboyantly attired with files and briefcases
and assorted aides,
cameras clicking, movies whirling,
the all pervasive air of tired concern,
the tinkering with this and that
lengthening into interminable hours—
statistics eddying up and down the table—
and then the exit,
not by the door wherein we went—
this one leads downstairs.

This was clearly written in a fairly cynical mood, along with another light poem called *The Seminar*, which I have found in several anthologies of Indo-Anglian poetry:

There sat the learned,
like rows of soda-water bottles
awaiting their turn to be drunk,
and one by one
they waddled to the mike and there unfurled
their private prejudices
garbed in the robe of reason.
One, a jaunty little cockroach,
leapt to and fro while speaking;
another stood in solemn grandeur
like a statue carved with moving lips;
a third, bearded and grim
and glaring like a hawk,
bore down in garrulous ferocity
upon the audience.

And thus it went
until, in God's own time,
their words sloshed all around the crowded hall,
they mercifully left for lunch.

In fairness I must say that this second poem did not describe the Governors' Conference but one of the many tedious seminars that I had to attend in Srinagar. Also I should add that only a small fraction of my poems were satirical.

After the Governors' Conference I went down to Bombay. The metal plate inserted into my left leg after the car accident in 1949 had begun giving trouble. My leg would swell up whenever I stood or walked for any length of time, and the doctors in London had advised that it should now be removed. I went into the Breach Candy Nursing Home for a few days, where the eminent orthopaedic surgeon, Dr K. T. Dholakia, performed the operation under general anaesthesia. While recovering from it there were two tragic occurrences. On 22 November a helicopter crashed near Poonch killing several senior officers well known to me. Among them were the Northern Army Commander General Daulet Singh, the Udhampur Corps Commander General Bikram Singh, and Air Vice Marshal Pinto, all outstanding officers with brilliant service records. Stunning as this disaster was, it was overshadowed by the shocking news the same day that President Kennedy had been assassinated.

Recently, twenty years after that tragic event, I read several assessments in *Time* and *Newsweek* about his presidency. Whatever the historians may say, it was for me a shattering moment. He was a brilliant and charismatic young leader of the world's mightiest democracy, who gave promise to many of my generation of a better world. He had mentioned Jawaharlal Nehru in his famous inaugural address, and had shown after many years of the 'Dulles approach' a fresh understanding of India and its problems. I dashed off an article from my hospital bed called 'The passing of President Kennedy; a personal memoir by one who never met him', which was published in the *Times of India*. I recalled the election day and wrote:

> He seemed to enshrine a promise for a brighter future, for a liberal, dynamic and creative policy which would help us all create a better world. . . . On the night of the presidential election I happened to

be in a suite at Rashtrapati Bhavan where I was attending the annual Governors' Conference. I stayed glued to the radio almost throughout the night. . . . The next morning we were in the midst of the conference when a messenger came in bearing a chit on a silver tray which he handed to the Prime Minister. He read it and interrupted the proceedings with the remark 'Kennedy has won the election'. It took all my power of self-control to prevent myself jumping up and dancing round the table. The Prime Minister was also visibly pleased.

In fact, Kennedy's assassination, followed closely by those of his brother Robert Kennedy and Martin Luther King, shattered the hopes of millions who had yearned for a better world. Taken together, these three killings did something to my perception of the United States; it has never been the same again. The era of hope had been terminated and the world had entered a new, dangerous phase where the dark forces of violence increasingly threaten to take over and destroy human civilization. It was certainly not without significance that these three men, all expertly assassinated, headed the hate-list of the malevolent J. Edgar Hoover, Director of the FBI.

A month later I was in Varanasi again to preside over the Convocation, when I got an urgent and startling message from Jammu. The holy relic of the Prophet Mohammad, enshrined for three centuries at the mosque of Hazratbal, had been found missing after the Friday prayers on 28 December. The news spread like wildfire throughout Srinagar and the entire valley. Within hours there was a sensation among the Muslims, whose religious sentiments were deeply outraged. The relic, consisting of a single hair from the beard of the prophet, had been an object of supreme veneration in Kashmir ever since it was installed there during the reign of the Moghuls. On special occasions during the year the relic, lodged in a small crystal tube, was brought out and shown to the faithful gathered at the shrine. I realized immediately that this was potentially a crisis of the gravest magnitude, and I boarded the first train to Pathankot, which was then the railhead for Jammu. I reached Jammu on the 31st morning and recorded a message for the radio expressing my deep anguish at the theft and associating myself with the sentiments of the Kashmiri people.

Shamsuddin had rushed up to Srinagar from Jammu immediately on hearing the news of the theft, and rang me up as soon as I reached Jammu. From what he said and from other sources, it became clear to me that the situation was rapidly getting out of hand. The pent-up resentment against Bakshi Ghulam Mohammad and his family had broken out into the open. Several shops and cinema houses belonging to them in the fashionable Residency area were attacked and burnt by mobs. The whole valley went into mourning; people wore black, and cars were allowed to ply only if they flew black flags. It was bitterly cold in Kashmir, but many people were so shocked that they stopped lighting their kitchen fires. Women came out onto the streets in black gowns crying and wailing, though luckily there was no communal tension. On the 28th morning some political leaders including Shafi Qureshi and Ghulam Mohiuddin Quarra, President of the pro-Pakistan Kashmir Political Conference, were arrested. The Director of the Intelligence Bureau, B. N. Mullick, had reached Srinagar from Delhi and had begun investigating the case with the assistance of the able Inspector-General of the State Police, Lakshman Das Thakur, and the Assistant Director of the Intelligence Bureau, Hamir Singh.

As soon as my fears regarding the gravity of the situation were confirmed, I flew back to Delhi by an Indian Air Force plane on 2 January, where we had a meeting in the Prime Minister's room in South Block. Among those present were the Home Minister, Gulzarilal Nanda, the Minister without Portfolio, Lal Bahadur Shastri, the Home Secretary, V. Viswanathan and the Cabinet Secretary, S. S. Khera. Jawaharlal was deeply worried at the crisis, and we discussed it in depth. Finally, it was decided that I would fly to Srinagar the next day accompanied by Viswanathan and a hand-picked team of officials. Efforts would be made to help the State government handle the situation but, if it began getting out of control, the ultimate option of imposing *Sadar-i-Riyasat's* rule under the State Constitution would have to be considered. In that eventuality, Viswanathan would act as my Adviser and his team of officers would help take over key assignments in the State administration.

On the 3rd our whole party, which included Viswanathan and several IAS officers, including P. K. Dave, Srinivas Vardhan and Sushital Banerjee, as well as a senior Muslim police officer from U.P., left by an Air Force Viscount for Srinagar. Due to inclement weather in the valley, we were forced to land in Jammu, and after waiting for about two hours, during which the weather showed no signs of improvement, we left for Srinagar by road at about noon. Our long convoy got into the city late that evening. It was bitterly cold, and instead of staying at Karan Mahal I had decided that it would be more convenient to stay with my cousin Nasib Chand who managed my orchards and lived in a snug house called Cherry Grove inside our main orchard facing the Dal Lake.

On the 4th morning my first action was to visit Hazratbal itself. I did not consult Shamsuddin, as I knew he would advise me against going on security and political considerations. However, I was quite clear on what I should do and, besides, I had genuine reverence for the shrine. I went in a private car without a flag, with only Nasib and a staff officer accompanying me. I did not allow either the pilot or the police escort to come, because any show of pomp or authority on that occasion would have been most inappropriate. As soon as we reached the shrine I was immediately mobbed by thousands of people who had been virtually camping there since the relic had disappeared a week earlier. They seemed genuinely delighted to see me, and pressed around me. Some of the elderly people kissed my hands and embraced me saying, '*Hamara Badshah aa gaya*' ('Our king has come'). I was deeply moved to see that, despite three decades of virulent anti-Dogra propaganda and brain-washing by Sheikh Abdullah and his party, the Kashmiris retained regard and deep affection for me even in their hour of anguish.

I proceeded into the shrine and offered prayers for the speedy return of the holy relic. I also directed that special prayers on my behalf be conducted at leading Muslim, Hindu and Sikh shrines in the valley and for this purpose sent eleven hundred rupees each to eleven such shrines. On returning from Hazratbal, I began a series of meetings and consultations with political leaders as well as senior civil and military officials. Among those I met were G. M. Sadiq, Syed Mir Qasim and D. P. Dhar, who urged me to dismiss the Shamsuddin government and promulgate *Sadar-i-Riyasat's* rule without further delay; an all-party deputation from Jammu, consisting of the Praja Parishad,

Praja Socialist Party and Akali Dal representatives, also favoured the same course of action. A fifteen-member Action Committee had been formed under the Chairmanship of Moulvi Mohammad Sayeed Masoodi and including Sheikh Abdullah's eldest son Sheikh Farooq (now better known as Dr Farooq Abdullah) and Mir Waiz Maulvi Farouq. The State administration had virtually come to a standstill, the National Conference was nowhere to be seen, and the Action Committee was in effect running an unauthorized parallel administration, controlling traffic, prices and commerce. I called for Bakshi Ghulam Mohammad, but he had left by road for Jammu on the previous night.

B. N. Mullick also met me along with L. D. Thakur and Hamir Singh. It was clear that Shamsuddin could not handle the situation, and that, if the holy relic was not recovered within the next 24 hours, there would be no alternative but to declare *Sadar-i-Riyasat's* rule. I told Mullick that I would give him another 24 hours before recommending to Jawaharlal that the Ministry be dismissed. Mullick, an able and deeply committed officer who was endowed with a self-will that almost bordered on arrogance, had been personally supervising the investigation of the theft. He disclosed that the windows of the *sanctum sanctorum*, where the relic had been lodged, were broken outwards, which clearly showed that it was an insider's job, or at least that one of the accomplices had worked from within the shrine. On the evening of 3 January he devised a brilliant strategy. It was made public that for the next 24 hours all guards around Hazratbal would be withdrawn, so that whoever had taken the relic could put it back without arousing suspicion. In my view it was the State Officer, L. D. Thakur, whose contacts made the crucial breakthrough.

Despite all the prayers and investigations, I was not sure whether the relic would in fact be recovered. I started to draft a message to the Prime Minister. I was sitting in my office at Karan Mahal at about six that evening when I got a telephone call from Mullick to say that the relic had been recovered, having been replaced from where it was stolen, and that he had ceremoniously taken it and placed it in the office of his Assistant Director. As soon as I heard the news it was as if a burden had been lifted from my shoulders. There was also great joy in the city, but many people were sceptical and the radio in Pakistan-occupied Kashmir poured out a constant stream of venom during this whole period. When the report of the recovery was announced they started saying that it was a fake and that the original relic had been replaced. Thus, with the recovery of the relic the crisis was by no means over.

That afternoon the Action Committee held a public meeting in Srinagar and raised five demands: that the relic should be identified by a panel of people known for their integrity and fearlessness; that it should be returned as early as possible to the Hazratbal shrine under the legal custody of responsible officials; that the investigation of the case should be taken over in its entirety by the Government of India; that the case against the culprits should be tried by a High Court Judge from outside the State; and that those arrested on the 28th evening in connection with the public demonstration should be released.

These demands were moderate and sensible, and despite the virulent propaganda from Pakistan in its press and radio ever since the disappearance of the relic, the Action Committee did not allow the movement to fall into anti-national hands. While their distrust of the Shamsuddin government was total, their appeal was not to any outside power but to the Government of India. In this, the role played by the veteran Maulana Masoodi was crucial and praiseworthy. A deputation led by him and including Dr Farooq Abdullah, Maulvi Farouq, Mufti Jalaluddin, Moulvi Mohd Yasin, Mohammad Abbas and Ghulam Rasool met me on 7 January and was with me for over two hours. They reiterated the points made at their public meeting, and elaborated on various aspects of them.

One new point that they raised was that Bakshi Ghulam Mohammad had issued a statement blaming 'anti-national elements' for the theft, and had returned to Srinagar so as to be able to influence State officials who were mainly his creation. They hinted that Bakshi or Rashid had in some way engineered the theft to prove their indispensability and thereby stage a political comeback. There was another theory circulating that Bakshi's mother, who was seriously ill in Jammu, had expressed a desire to have *darshan* of the relic before she died, and that the relic had been surreptitiously removed for that purpose. Mullick, on the other hand, was convinced that the whole episode had been master-minded by Pakistani intelligence with the intention of creating chaos in the valley and ultimately seizing this long-cherished prize.

On 3 January, Sheikh Abdullah wrote a long letter to the President, Dr Radhakrishnan, from the special jail in Jammu in which he linked the theft to what he called the process of 'de-humanization' that was initiated with his arrest in 1953. He also hinted at political motivation, and wrote, obviously referring to Bakshi, 'It is quite possible that the motive behind this sacrilege is purely political. Whoever had the brute courage to do it, he is likely to thwart and defeat all inquiries and investigations.' He concluded by calling for a revision of the Kashmir policy since 1953, which he described as 'the root cause of all the evils, which culminated in the present tragedy'.

G. M. Sadiq also wrote a long letter to the Prime Minister on 15 January 1964; in it he alleged that Bakshi and his supporters had done their utmost to communalize the situation and encourage the use of force against the demonstrators, which would have resulted in disaster. It was an important document, in which he attacked many of the established beliefs that had developed during the ten years of Bakshi's rule. He wrote:

> One of the beliefs which has been commonly entertained in the past is that the influence of Pakistan on the Kashmir Muslims is fairly wide and firmly rooted. From this belief has stemmed a primordial fear of the people. Based on this fear, policies have developed and influenced every sphere of our life. . . . The rule of law was thrown overboard and all executive agencies as well as the judiciary were bent to the service of a clique in the National Conference. Corruption in all branches of administration and political life was considered to be something inevitable in this situation. . . . In this the Bakshi family held a place of significance, absolute prominence and power. From all this a myth was born that the National Conference is a sturdy organization of the people headed by two strong men, the twins of the Bakshi family. During the last ten fateful days the people of Kashmir, every man, woman and child irrespective of religion or party affiliations has given a lie to these assumptions. When the whole population of the State was on the move, these elements of the National Conference who had harmed the interests of the people for their personal benefits were hiding in their rat holes.

He went on to add:

> The other noteworthy feature of the situation was the heart-warming scenes of communal amity; and for the first time in the history of the State the presence of complete unity in its geographical constituents. The Hindus, Sikhs and Muslims mingled together and the blood of their common ancestors flowed in their veins, and tears came into my eyes when I heard their full throated cries of *Har Har Mahadev*, *Allah-o-Akbar* and *Sat-Siri-Akal*. This was the greatest moment of glory in their history. In this demonstration of civilized behaviour who would not be proud to be a Kashmiri. In forging this unity between the people of Jammu and of Kashmir the credit goes to the saner sections of the National Conference in Jammu and more prominently to Pandit P. N. Dogra. He along with others successfully combated the rumours of communal riots in Kashmir assiduously spread by the pro-Bakshi sections of the National Conference in Jammu.

Further, Sadiq pointed out how the religious factor had been distorted and exploited. With his characteristic clarity of mind and intellectual integrity he wrote:

> We have been too often led to believe that anything that smacked of the Hindu world injured the sentiments of the Muslims here. Therefore, we have been asked to prefer and tolerate a Muslim politician or a Muslim civil servant even though he be corrupt and tyrannous. . . . Why is it that Dr Karan Singh, a scion of the Dogra Dynasty, was mentioned with the deepest affection and respect in the public meetings attended by lacs of people? Why is it that Sheikh Sahib's son and many other speakers from public platform demanded and continue to demand the rule of Sadar-i-Riyasat in preference to the rule of the 'Muslim' Bakshi? The people of Kashmir have torn to pieces shibboleths in which, alas, the Government of India has been led to place its faith by interested and self-seeking individuals for many years in the past.

I have quoted at some length from this letter because, after the collapse of the political apparatus built by Bakshi since 1953, Sadiq represented the only viable political force in the State capable of picking up the pieces. I had throughout kept in close touch with him, Mir Qasim and D. P. Dhar who, between them, dominated this group. The political expression of this was the emergence of the Congress led by this trio as the new political force in the State. As Sadiq put it bluntly in the same letter:

Another lesson which we must boldly draw from the situation is that the emergence of the Congress in the State as a prelude to closer political relations between Kashmir and the rest of the country is a matter of historical necessity. National Conference no doubt was a powerful organization with a glowing past. It is the traditions fostered by this which are responsible for the sanity and goodness of our people. But that National Conference is dead. What we are trying to do now is to worship a stinking carcass.

In my detailed political assessment to Jawaharlal I laid stress on this aspect, and urged that, while the National Conference should now pass into the pages of history, a bold political move should be made by launching the Indian National Congress with the Sadiq group plus a broad spectrum of politically aware and enlightened sections of the State. This could be done either by instituting an interregnum of *Sadar-i-Riyasat's* rule or by persuading Shamsuddin to resign.

It was at this stage that Lal Bahadur Shastri was entrusted by the Prime Minister with the task of sorting out the political problems in the State. The first question that had to be faced was the thorny matter of verifying the authenticity of the relic. Pakistani propaganda apart, the strange circumstances of its disappearance and recovery had caused grave misgivings in the minds of the people, which had to be allayed if normalcy was to be restored. I went to Delhi again and had prolonged discussions with Lal Bahadur. He flew up to Srinagar on 30 January on a three-day visit during which, with his characteristic patience and strong commonsense, he worked out a mutually acceptable formula for identification of the holy relic by Muslim divines before the public *deedar* fixed on the 21st day of Ramzan, corresponding to 6 February, despite B. N. Mullick's stiff opposition to the whole idea. The identification was held under dramatic circumstances by a band of Muslim divines led by the saintly octogenarian, Miraq Shah, whose integrity was beyond dispute. In a packed hall in the presence of Lal Bahadur, the sage slowly raised the crystal tube to the light. For a long moment he peered at the tiny hair suspended within it, while everyone in the hall held his breath. Then in a firm voice he announced, '*Al Haq*' ('it is real'). Kashmir had literally been saved by a hair's breadth.

The next step was somehow to ease Shamsuddin out of office and install Sadiq in his place. Bakshi Ghulam Mohammad, after remaining virtually underground during those turbulent days, suddenly became vociferous in supporting the Shamsuddin government and threatening dire political consequences if he was dismissed. Meanwhile, the Home Minister, Gulzarilal Nanda, announced that the three persons connected with the theft of the holy relic were Abdul Rahim Bandey, Abdul Rashid and Kabir Butt, of which the first was a *Mutwalli* or priest of the shrine. He also announced that a judge from outside the State would try the case. In Jammu, the Praja Parishad and other opposition parties had made elaborate arrangements to observe a complete strike on 19 February to express their lack of confidence in the government, but I was able to persuade them to postpone their protest because of Lal Bahadur's impending visit. He came on the 20th and spent three days in Jammu during which he had discussions with political leaders of all persuasions.

It is significant that, although Gulzarilal Nanda was Home Minister, it was Lal Bahadur to whom Jawaharlal entrusted the delicate political task of tackling Jammu and Kashmir. Indeed, after his illness Jawaharlal had come to depend increasingly upon Lal Bahadur, and had recalled him to the Cabinet as Minister without Portfolio. Gulzarilal was a devout and well-meaning person, but as an administrator he was ineffectual and he also lacked Lal Bahadur's grass-roots support in the Congress Party. This fact was to become clear later that year.

Lal Bahadur persuaded the National Conference to hold a meeting to elect a new leader on the 28th. Bakshi Ghulam Mohammad left for Delhi immediately after Lal Bahadur, and had a meeting with Jawaharlal at which he was told in no uncertain terms that he must co-operate in the change of leadership or face the consequences. He returned to Jammu on the 26th, and on the 28th morning the National Conference Parliamentary Party met. Bakshi proposed the name of G. M. Sadiq who was unanimously elected leader, following which Shamsuddin tendered his resignation. I called upon Sadiq to form a new government, and that evening administered the oath of office to a four-man Cabinet led by him and including Mir Qasim, D. P. Dhar and Trilochan Dutta. Lal Bahadur was present at the swearing-in ceremony, which was held at our picturesque residence Hari Niwas.

Thus, after a stint of slightly over four months, the ill-starred Shamsuddin Ministry passed into oblivion. It had never really been a viable political entity, because it was manipulated from outside by the Bakshis. Shamsuddin in fact was a rather pitiable figure. With the advent of Sadiq the reins of office passed into firm hands. A man of integrity and political stature, Sadiq was inducted at a crucial juncture. This was just as well, because within eighteen months the next major crisis was upon us.

Chapter 21

The traumatic episode of the holy relic forced Jawaharlal to reconsider the whole structure of his Kashmir policy. He had always been unhappy about Sheikh Abdullah having to be detained, and with Sadiq in the saddle he felt that the time had come to make a move towards Sheikh Abdullah, now that his *bête noire*, Bakshi Ghulam Mohammad, was no longer in power. Abdullah, understandably, was bitter to the core as far as Bakshi was concerned, and in fact carried this antipathy even beyond the grave by refusing to let Bakshi be buried near the Jama Masjid. As long as Bakshi was in power, a rapprochement with Abdullah was not really possible, and the one attempt made when releasing him from detention did not last more than a few months. Thereafter an elaborate case was made out against Sheikh Abdullah called the Kashmir conspiracy case which went on for several years.

Sadiq's policy of 'liberalization' also involved, as a logical corollary, a change in the government's attitude *vis-à-vis* Abdullah. Quite soon after he assumed office, he had detailed consultations with the Prime Minister as a result of which there was a major policy shift and a decision taken to withdraw the conspiracy case against Abdullah, Afzal Beg and others and to release them unconditionally. An application for withdrawal was made on 8 April 1964 before the court of the Additional Sessions Judge who was hearing the case, and Sheikh Abdullah and his colleagues were released the same day. Describing Sadiq as 'an old comrade' in his press conference the next morning, Abdullah said that, while it was the people of Jammu and Kashmir who had to decide their future, he had no ready-made solution for the Kashmir problem. He did, however, say that he was definitely against the division of the State.

Immediately after the press conference, Sheikh Abdullah, accompanied by Mirza Afzal Beg, called on me at Hari Niwas. It was our first meeting after the memorable confrontation on 8 August 1953 over a decade earlier. He seemed to have lost none of his vigour, and we embraced cordially and chatted for half an hour without touching on politics. His call was evidently a gesture of goodwill and reconciliation and, although I had grave reservations regarding the new policy, I felt that it should be given a fair trial. In my next letter to the President dated 9 April 1964 I mentioned some of these doubts. I wrote:

It is too early to assess the full political implications of Sheikh Abdullah's release. An important part of his attention may get diverted towards strengthening communal harmony in the country, which of course will be a laudable endeavour for which he will deserve praise. The crucial situation, however, will arise when his attitude towards the political status of Jammu and Kashmir State becomes clear. Is he prepared to accept not only the accession of this State to India but also its coming on par with the other States of the Union, a process popularly but mistakenly known as 'integration'? From the trend of his press conference this does not appear likely. Will he support a 'limited' accession on the basis of the original Instrument on the three subjects of Defence, Foreign Affairs and Communications? This would involve turning back the clock and undoing much that has been done over the last eleven years since he was first detained. Will he for all practical purposes reject the accession altogether and come out in favour of an independent Kashmir to act as a 'bridge of understanding' between India and Pakistan? These questions remain to be clearly answered, and upon his attitude to them will depend to a large extent his political future. At present, despite genuine apprehensions and fears among various sections, there is the inclination to 'wait and see', to give him time to re-establish his contacts before he comes to a political decision.

Pakistan of course was jubilant at Abdullah's release, though the jubilation was not unqualified because they knew him well enough to realize that he was far too significant a political personality to act as anyone's pawn. Bhutto invited him to meet President Ayub, and when questioned about this at his press conference Abdullah said, 'Let us wait and watch. This is a matter between the two Governments [India and Pakistan]. This may arise when I meet Nehru.' After touring the Muslim-majority Doda district of Jammu, Abdullah entered the valley on 17 April and received a tumultuous welcome from Anantnag all the way to Srinagar and, indeed, wherever else he went in Kashmir. There was never any doubt that he was the most popular and charismatic Kashmiri leader. His speeches, however, were ambiguous. While on the positive side he laid great stress on the maintenance of communal harmony and urged the Kashmiri Muslims to safeguard the life and property of non-Muslims, as far as the political angle was concerned it did not seem that he had appreciated the changes in the situation since 1953. He refused to accept that the accession of the State to India was permanent and final, he came out openly in favour of the Plebiscite Front which had been founded by his main lieutenant Afzal Beg, and he seemed to equate India and Pakistan by reiterating that a solution 'of the Kashmir dispute should be found which is acceptable to all three parties'. His tour and speeches created a tremendous impact in the valley.

In Jammu, of course, they had the reverse reaction. Jammu had given Abdullah a warm and generous reception after his release in the hope that he would be encouraged to accept the new realities. His stance caused considerable dismay in Jammu, even among his small band of supporters. Once again, the mutually contradictory forces built into the structure of Jammu and Kashmir became obvious. What was manna in Kashmir was anathema in Jammu, and vice versa. This ambiguity virtually ensured political instability in the State, which continues down to the present day. Sadiq and his colleagues faced a difficult situation with equanimity. Sadiq himself unequivocally reiterated the permanence of accession, rejected Abdullah's stand regarding 'self-determination' and reaffirmed that law and order in the State would be maintained. The role of Bakshi Ghulam Mohammad was ambiguous, to say the least. He met Abdullah, and our reports were that he tried to convince him that he would have released him much earlier had it not been for the intervention of the Government of India. He met him at Mujahid Manzil, the old National Conference headquarters in Srinagar, ostensibly to discuss affairs relating to the Muslim Auqaf Trust. When he and his colleagues emerged they were showered with abuse and stones by the large crowd that had gathered outside during the meeting.

The whole political situation in effect had become destabilized, but this was part of the calculated risk that had been taken in releasing Abdullah, what Dr Radhakrishnan had called 'an act of faith'. It was clear to me that either Sadiq and his group had to capture full control of the National Conference, or they must start a new party. A situation in which the masses were with Abdullah, the party machine with Bakshi and the government with Sadiq was clearly untenable. The other political group in the valley—the Action Committee led by Maulana Masoodi, the Political Conference led by Mohiuddin Quarra and the elements led by Maulvi Farouq—were also overshadowed by Abdullah, though they retained their separate entities.

All eyes were now turned to Delhi, which Sheikh Abdullah was scheduled to visit for talks with Jawaharlal. As I wrote in my fortnightly letter of 24 April: 'There would appear to be little doubt that if the Sheikh continues on his present path he is bound sooner or later to come into conflict with the legally established Government. Sadiq's policy is to avoid any clash as far as possible, while firmly maintaining law and order. But if the Delhi talks do not produce the desired modification in the Sheikh's attitude a showdown is inevitable with all the grave consequences that will be implied therein.'

In order to brief the leaders at Delhi on the situation in detail, I thought I should go down myself. I spent a fortnight there from the end of April to 11 May, during which time I met the Prime Minister, Gulzarilal Nanda and Lal Bahadur Shastri. Jawaharlal had suffered a stroke in Bhubaneswar on 6 January from which he never really recovered. His last letter to me was a brief note giving me an appointment on 10 May at 12.30 P.M. When Asha and I entered his room in Teen Murti House he was seated in a chair and made an effort to rise. When I requested him to remain seated he said with a smile, 'I have not forgotten my manners', and got up. Clearly, he was seriously ill, and it was only his indomitable will that kept him going until the very last day. I inquired after his health, but did not feel it appropriate to engage him in a long conversation about the situation in Kashmir. I simply said that I had discussed the situation in detail with Lal Bahadur. He smiled when we took leave of him, and I bent and took his hand in mine. It was the last time I was to see him alive.

Sheikh Abdullah also came to Delhi at the end of April and had an emotional meeting with Jawaharlal, with whom he stayed at Teen Murti House as his personal guest. He had prolonged meetings with the Prime Minister as well as a number of Cabinet Ministers and other political leaders, including

Jayaprakash Narayan, who had always been friendly and favourably disposed to him. In fact, Abdullah and Jayaprakash both called on me separately in Delhi. Abdullah succeeded in persuading Jawaharlal to let him have one more try at building a bridge between India and Pakistan. Jawaharlal was no doubt disturbed at the growing rift between the two countries, and his thinking had obviously been influenced by the ill-fated border war with China. Also, the USA and Britain had kept up their strong pressure on him to come to some sort of an agreement with Pakistan, and perhaps he saw in Sheikh Abdullah a person who might possibly achieve this and also relieve India of the constant pressure *vis-à-vis* Kashmir.

I flew to Srinagar on 11 May. On the 12th Sheikh Abdullah issued a statement in which he mentioned his various meetings and discussions, reiterated the importance of India and Pakistan coming to an agreement over Kashmir, and announced that he was visiting Pakistan shortly to have discussions with President Ayub Khan and other leaders there. Jawaharlal also made statements in Parliament and at the AICC supporting these moves. Speaking at the AICC he said:

> Sheikh Abdullah is wedded to the principles of secularism and does not want anything to be done to vitiate these in any way. He does not believe in the two-nation theory which was the basis of formation of Pakistan. Nevertheless, he hopes that it should be possible for India, holding on to her principles, to live in peace and friendship with Pakistan and thus incidentally to put an end to the question of Kashmir. I cannot say if we shall succeed in this, but it is clear that unless we succeed India will carry the burden of a continuing conflict with Pakistan with all that this implies. I hope Pakistan will get rid of its hatred and fear of India. She has nothing to fear from India, unless she herself attacks India. I hope that it may be possible for the two countries to develop closer and more intimate relations to the advantage of both. If Sheikh Abdullah can help in bringing this about, he will have done a great service to both the countries. We are prepared to help him in this attempt, but in doing so we must adhere to our principles as well as our basic attitude in regard to Kashmir.

This marked a dramatic change in Jawaharlal's policy *vis-à-vis* Kashmir. Only a few weeks earlier Sheikh Abdullah was in the dock, being accused of conspiring with a foreign power to overthrow the lawfully established government of the State. To put it bluntly, he was being tried for treason. Now, with the blessings of the Prime Minister, he was embarking on a visit to Pakistan with a view to finding a mutually acceptable agreement over the State of Jammu and Kashmir. This curious dichotomy well illustrates the basic ambivalence that has characterized our approach ever since 1947. We claim the whole State ruled by my father as Indian territory, and yet we do nothing to recover the vast areas under foreign occupation. Even when we reoccupy some territory, as we did in 1965 and 1971, we still refrain from clinching the issue and cloud the entire affair under elaborate clichés of 'bilateralism' and substituting 'line of actual control' for 'Ceasefire Line'. We thus neither cut our losses and settle the issue, nor clinch the matter once and for all by virtue of our superior military power. The beautiful State founded by my ancestors, therefore, seems destined to remain an apple of discord, its people sentenced to permanent instability.

Abdullah returned to Srinagar from Delhi on 13 May and addressed a public meeting that afternoon. His tone was more balanced and moderate than in his earlier speeches. He reiterated the importance of safeguarding the interests of the minorities both in India and Pakistan, and counselled patience while negotiations were in progress. His emphasis shifted from 'self-determination' to Indo-Pakistan reconciliation, and he announced that he would leave for Delhi on 21 May *en route* to Pakistan. He reached Pakistan on the 25th to a tumultuous welcome, had talks with President Ayub Khan and Bhutto that day and the next. On 27 May he reached Muzaffarabad, the capital of what Pakistan called 'Azad Kashmir' and what India has consistently termed 'Pakistan-occupied Kashmir'. What the outcome of his visit would have been is something that must for ever remain a matter of speculation, because on 27 May an event occurred that radically changed the whole situation.

Ever since he suffered a stroke in Bhubaneswar, the Prime Minister had been a dying man. It was a tribute to his intense dedication that, despite this, he kept working till the very end. After the meeting with Abdullah he went to the Indo-Nepal border for a function at a village called Bhainsalotan. Returning to Delhi, he went off again to Dehra Dun and stayed in the Circuit House there with its beautiful lawns and flower beds that he so loved. He came back to Delhi again on 26 May and worked at his desk till late at night, clearing all the papers as was his habit so that not a single file was pending. On the morning of the 27th he suffered a major stroke. We received the news in Srinagar almost at once, and in a few hours our worst fears had been confirmed. Jawaharlal Nehru, patriot and politician, historian and litterateur, administrator and world statesman, was no more.

From the very first day of independent India, and, indeed, long before that as head of the Interim government and intrepid freedom fighter, he had laboured day and night for the welfare of the nation. Travelling to every nook and corner of the country, and carrying the message of peace and non-alignment to the four corners of the world, he had worked incessantly for India's progress and the welfare of humanity. Rarely in history has a leader of a freedom struggle lived to preside for almost two decades over the destiny of his nation. Jawaharlal was a unique leader in many ways. Socialist and revolutionary by nature, he yet imbibed the spirit of his mentor Mahatma Gandhi and tempered his radicalism with a realism that was born of intimate contact with his people. Born to privilege, he mingled with the masses of India and conversed with them in a manner few have ever rivalled anywhere in the world. While his English was chiselled and polished, reflecting his elitist upbringing, his Hindi was down to earth and simple, a language to which the common people of India responded in unique measure. Despite his impatience and bursts of temper from time to time, his patriotism and sincerity were so patent that even his opponents were disarmed. Winston Churchill once summed this up in a short remark to his sister, Vijayalakshmi Pandit. 'Jawaharlal', he said, 'had no fear, and he had no hate.'

For me, Jawaharlal, 'Panditji' as we always called him, was a major factor in my public career. In *Heir Apparent* I have called him my 'political guru', and that indeed he was, because it was first from his books and then from himself that I was initiated into the Indian political process. This is not to say that I agreed with his approaches to all issues; no *shishya* worth his salt is expected to do that. In particular, perhaps because of his European upbringing, I found that his approach to religion in general and Hinduism in particular was unsatisfactory. But the fact remains that he was the major political figure in my life, and I owe an immense debt to him for the affection and kindness that he consistently displayed towards me from 1949, when at the age of eighteen I became Regent of Jammu and Kashmir on his intervention, right down to his passing away fifteen years later.

As soon as we heard the news I hurried to the airport where an Air Force plane had been requisitioned to fly us to Delhi. D. P. Dhar was there in tears, and we embraced wordlessly. Sadiq also flew with us to attend the funeral, for which Sheikh Abdullah had rushed back from Pakistan. Thanks to bad weather, which always seemed to dog our footsteps at critical junctures, we were not able to fly down on the 27th. We left early the next morning and got to Delhi well before the funeral procession began. A sea of humanity stretched all the way from Teen Murti to Rajghat near which Panditji was cremated. Not surprisingly, it was the biggest funeral procession since Gandhiji was assassinated in 1948. Millions flocked to have a last glimpse of their beloved leader. Indira Gandhi, Rajiv and Sanjay accompanied the cortege, as did other members of the family.

For the funeral itself, leaders from all over the world had assembled. I saw Zulfikar Ali Bhutto for the first and last time on that occasion. In accordance with Panditji's last will and testament, an eloquent and moving document, his ashes were to be scattered throughout the length and breadth of India. I was reminded of my father's will about his own ashes, which were scattered over Jammu as I have recorded. Jawaharlal's ashes were placed in a large number of urns and carried to different parts of India. I took one urn and left by train for Pathankot on 4 June. Thousands had gathered there to pay homage to Panditji, and from the railway station I brought the urn in an open car seventy-eight miles to Jammu. People had gathered throughout the route, and at Jammu there was a very large turnout. The ashes were kept on a special rostrum at the Parade ground until they were immersed in the Tawi on the 8th.

Sadiq had flown up to Srinagar with a handful of the ashes on the 6th. I joined him there, and two days later we drove with them in an open car from Rajgadh to Shadipur. There we boarded a boat painted white and rowed by white-uniformed oarsmen, the very boat which had on so many occasions taken Panditji up the Jhelum in joyous river processions while the citizens of Srinagar lined both sides cheering and waving. He always had a special attraction for Kashmir, the land of his forefathers, and it was significant that till the very end of his unique and inspired life he was attempting to solve the Kashmir imbroglio. We immersed the ashes at the confluence of the Jhelum and its tributary the Sindh, because a *sangam* between two rivers is always considered particularly auspicious for the Hindus. Kushak Bakula took some ashes to Leh and immersed them in the Indus, the river from which our country originally gained its name. Thus ended the life of one of India's most distinguished patriots, and Jawaharlal passed for ever into the pages of history that he had chronicled with such eloquence, and in the shaping of which he had played such a noble and memorable part.

Ever since the Chinese debacle, the question 'After Nehru Who?' had become a favourite topic of discussion in India and, indeed, all over the world. It was quite clear to me that after the Kamaraj Plan, which knocked Morarji Desai out of the race, and the recall of Lal Bahadur Shastri after Jawaharlal's illness, he was the man who would succeed to the Prime Ministership. Apart from Jawaharlal himself having shown his clear preference by singling him out for recall from among those who left the Cabinet under the Kamaraj Plan, Lal Bahadur had excellent grass-roots support in the party, of which he had been General Secretary for many years. His deceptively unobtrusive style, combined with a shrewdness and robust commonsense, made him the obvious successor. In the event, after a brief interlude when the office was held by Gulzarilal Nanda, who ranked number two in the Cabinet, Lal Bahadur was unanimously elected leader of the Congress Parliamentary Party and became Prime Minister.

It took the nation some time to get over the emotional shock of Jawaharlal's death. We had grown so used to his colossal presence that it was difficult to imagine an India without him. And yet the vitality of our democratic system was such that it ensured a smooth succession. I made a memorial broadcast from Srinagar on 9 June, when the thirteen-day national mourning came to an end. On the 11th, Sheikh Abdullah addressed a public meeting organized by the Plebiscite Front, during which he broke down several times when referring to Jawaharlal. He said that his talks with President Ayub Khan had gone well, and that had Panditji lived the two leaders would have met shortly. He hoped that Lal Bahadur and Ayub would pick up the threads at the forthcoming Prime Ministers' Conference in London. He did not reveal details of his talks with the Pakistani President but, as far as I was able to gather, some sort of 'confederation' proposal between India and Pakistan, whereby Kashmir would enjoy a special intermediate position, was the line upon which his mind was working.

In my first letter to the President after Jawaharlal's death, copies of which went to the Prime Minister and the Home Minister, I suggested that the time was ripe to transform the National Conference into the Indian National Congress, expand the four-man Ministry led by Sadiq and take effective steps to tone up the administrative machinery in the State. This was because rapid political developments were clearly round the corner. Within a fortnight Maulvi Farouq, who had become President of the Action Committee and belonged to the Mir Waiz family that had traditionally been opposed to Sheikh Abdullah and his party, was ousted and formed a new political party called the Awami Action Committee. Sheikh Abdullah started a vigorous tour of the valley in which he called upon all those who were pledged to secure self-determination for the Kashmiris to unite under the banner of the Plebiscite Front, the only organization that had made sacrifices to achieve this end. With his amazing facility for shifting the political emphasis as and when the occasion demanded, he began reverting to his earlier, more strident, attitude on crucial political issues. The excuse his friends trotted out was that he had to do this in order to pre-empt Maulvi Farouq but, as I wrote to the President, 'if he begins himself to take a blatantly communal and extremist line, the distinction between the two will hardly retain much significance'.

During his visit to Delhi, Abdullah had announced grandiloquently that he would be visiting London to be available for consultations at the Commonwealth Prime Ministers' Conference, but the hostile press reaction that this evoked, along with the challenge from Maulvi Farouq in the valley, made him cancel his plan. An ominous development that had begun some months earlier was the increasing incidents on the Ceasefire Line from the Pakistani side, especially in the Poonch–Rajouri sector. Thirty such incursions were reported by the State intelligence service in the month of June alone. Despite President Ayub's declared policy of seeking friendship with India, it was clear that Pakistan was keeping all its options open. The new situation created on the one hand by the release of Sheikh Abdullah and on the other by the passing away of Jawaharlal Nehru evidently gave that country renewed hope that, at long last, an opportunity was opening up to get Kashmir, by agreement if possible, by force if necessary.

My own reaction to all these events was one of dismay. Having been committed ever since I assumed the headship of the State in 1949 to do everything possible to strengthen the integration of the State with the rest of India, I was aghast at the way events seemed to be developing in exactly the opposite direction. The only hope I could see lay in forging closer political ties between the State and the Centre, and there my oft-repeated pleas for the metamorphosis of the National Conference into the Congress at last began bearing fruit. Sadiq and his colleagues took the initiative, and despite strong opposition from the Bakshi group within the party and Sheikh Abdullah from outside, he finally made a substantial move in that direction. At a series of meetings of the National Conference Working Committee and Parliamentary Party in July, it was finally decided to authorize Sadiq to extend the scope of the Indian National Congress to the State. Mir Qasim resigned from the government to assume the General Secretaryship of the National Conference and pave the way for its eventual merger with the Congress. Thus, I thought, what we lost on the political swings could to some extent be regained on the party roundabouts.

Chapter 22

While all these hectic political events were in progress, our family life was also developing. Jyoti was growing up into an adorable if rather pampered girl, being the only child at the time. In the six years after her birth Asha had a series of abortions, so my father died without seeing a grandson and my mother was depressed that her insistent prayers for one had not been answered. We were therefore delighted when Asha conceived again early in 1964. The delivery was due in August, and we decided to go down to Bombay again, because, although my father had passed away, Dr Shirodkar was still active. In fact, his fame as a gynaecologist and surgeon had spread the world over, and there was even a special operation named after him.

We went to Bombay in July, where we stayed at Shalimar Hotel near Kemp's Corner, not because of the Kashmiri name but because it was conveniently close to the Breach Candy Nursing Home where the delivery would take place, and also quite near Dr Shirodkar's own clinic. The Shalimar is a small hotel, so we took a whole floor consisting of four suites, a pale echo of the Martinez in Cannes where I was born, but not too bad considering that over three decades had since elapsed. Asha and I, Jyoti and her friend Ruchi Jamwal, a niece of one of my childhood companions Digby, some servants and a staff officer constituted our party. Asha's youngest sister Bharati and her stepmother also came to Bombay for the occasion. Labour pains started late on the evening of 4 August, and early on the fifth morning our first son was born. We had already decided to call him Vikramaditya.

The birth of a son, at least in eastern civilizations, does have a special impact despite all that we may say about women's lib. My mother, to whom I telephoned the news from the Nursing Home, was overjoyed, and when the birth was reported in the papers the next day we were inundated with thousands of telegrams so that the Bombay postal circle had to get a special stamp printed saying 'Shalimar Hotel'. We returned in triumph to Jammu, where we received an affectionate greeting from vast crowds who had thronged to felicitate us all the way from Pathankot. Despite the 'abolition' of monarchy, the warm sentiments of the people were still clearly evident. I had never seen my mother so utterly delighted. Despite her poor health she personally distributed lavish gifts to all our servants, staff and her friends, and had extensive thanksgiving prayers offered at numerous places of religious worship.

I myself had inwardly committed three major offerings of eleven thousand rupees each—to our family temple Sri Raghunath Mandir at Jammu; to the Harmander Sahib at Amritsar and to the Hazratbal shrine at Srinagar. The rise to power of our family in the nineteenth century was due to the patronage and support of the great Sikh ruler Maharaja Ranjit Singh, and I have always acknowledged our debt to the Golden Temple, holiest of Sikh shrines. Similarly, with our Kashmir connection, I have had reverence for the most venerated Muslim shrine in the valley. As a Hindu I follow the great teaching of the Vedanta which looks upon all religions as different paths leading to the same goal.

While returning from Bombay to Jammu we stopped for a few days in Delhi, where the first meeting of the Nehru Memorial Committee, which was later converted into the Jawaharlal Nehru Memorial Fund, was held. Sheikh Abdullah was also a member, and he called upon me again at Mansarovar, our Delhi house. We had a long conversation, and although he asserted that he was not committed to any particular solution but, rather, was anxious to find some way out of the difficulty, it was clear to me that what he really favoured was some sort of semi-independent status for the valley guaranteed both by India and Pakistan. Rather surprisingly, he added a personal appeal to me to take more initiative and to play a 'positive' role in solving the problem of our disputed State.

I returned to Srinagar at the end of August, to find that the Plebiscite Front had stepped up its activities. 9 August, the date upon which Sheikh Abdullah had been arrested eleven years earlier, was observed as a 'Black Day' by the Front with a general strike and a huge procession in which Abdullah participated and which he later addressed. This time his speech had acquired markedly anti-Indian overtones. He launched a frontal attack on Indian leaders for their involvement in the 1953 events, and lashed out not only at Bakshi but also at Sadiq, Mir Qasim and D. P. Dhar, whom he accused of being arch conspirators against him. Thus, for the first time, Abdullah took a hostile stance *vis-à-vis* the Sadiq government. The Shastri–Ayub meeting had not materialized, and he was becoming restive and frustrated. It became clear to me then that he was once again on a collision course. Representing as he did what may be called chauvinistic Kashmiri sub-nationalism, and endowed with an ego to match his impressive physical appearance, Abdullah was not yet in a mood to accept the reality of Jammu and Kashmir being part of India. He was still stuck with the idea that Pakistan must in some way be brought into the situation to guarantee Kashmiri rights. It was only with the break-up of Pakistan after the 1971 war that he began to come to terms with what were for him the harsh realities of the situation.

Meanwhile, moves for conversion of the National Conference into the Congress also progressed. Conventions were held both in Jammu and Srinagar, attended by thousands of National Conference workers. With Bakshi out of power and unable to distribute largesse any longer, it appeared as if he had begun to lose his grip over the organization, with the exception of a die-hard core of legislators, while Sadiq and Qasim strengthened theirs. Shafi Qureshi, who had founded the Kashmir National Congress the previous year, decided to join the National Conference along with his workers and we thought this would also strengthen Sadiq's position. In politics, however, appearances are often deceptive.

In the opposition camp, the confrontation between the Abdullah and Maulvi Farouq groups grew fiercer. Known as the *shers* (lions) and *bakras* (goats) respectively, these groups had been at loggerheads ever since Abdullah launched his movement in the thirties. There were scuffles and clashes between the rival supporters, and in their public speeches the two leaders attacked each other fiercely—Abdullah accused the Maulvi of adventurism and irresponsibility, while the latter hit back with the charge that Abdullah was splitting the monolithic unity that had been forged among the Muslims after the holy relic episode. Within Abdullah's camp two discernible groups also began to surface, the 'hard-liners' led by Afzal Beg and Ghulam Rasool Kochak advocating a strong anti-Indian line, and the other led by Maulana Masoodi and Khwaja Mubarak Shah advocating a more reasonable stance.

A session of the Legislative Assembly was due in September. I was away in Delhi for a week in connection with a series of meetings, including that of the Jawaharlal Nehru Memorial Fund. On my return to Srinagar I was horrified to discover that while Sadiq and his group of intellectual supporters were blissfully unaware of what was happening, Bakshi Ghulam Mohammad, with his characteristic boldness, had decided to make a determined bid to topple the government on the floor of the house. The action taken against his supporters and the threat of an inquiry into allegations of corruption against him had driven him to the wall. He called on me on 21 August and said quite clearly that he was determined to push ahead with the vote of no-confidence and was quite prepared to face any consequences that might be involved. He complained bitterly about the campaign of 'victimization and vilification' launched against him by the Sadiq government, and said that he was no longer prepared to put up with this. According to him, Sadiq, Qasim and D. P. Dhar were no less corrupt, and he produced a statement detailing their transgressions.

I tried to persuade Bakshi against taking a rash step, but he was angry and in no mood to listen to reason. Simultaneously, we learnt that he had succeeded in getting the signatures of a majority of legislators for the no-confidence motion, which was not surprising considering the fact that nearly two-thirds of them had been returned unopposed, almost all handpicked by Bakshi and owing their political allegiance to him. This placed us in an acute dilemma. With Abdullah on the warpath in the valley, we simply could not afford to have the Sadiq government toppled, particularly as the Kashmiris had shown their anger and resentment at the Bakshi group during the holy relic agitation. If the no-confidence motion went through, there would be no alternative but to place the State under *Sadar-i-Riyasat's* rule which, under the circumstances, would have been politically exploited to the hilt by Abdullah and the Plebiscite Front.

I immediately went into consultation with Sadiq, Mir

Qasim, D. P. Dhar and Shankar Prasada, who was then Secretary for Kashmir Affairs. We debated in my office at Karan Mahal deep into the night, and after prolonged consideration decided that the lesser evil would be to arrest Bakshi and some of his main supporters, and prorogue the Assembly. While this was quite clearly not a democratic move, we were really left without any viable alternative. The arrests were made the next morning under the Defence of India Rules, and Bakshi Ghulam Mohammad was in turn taken for detention to Tara Niwas in Udhampur where eleven years earlier we had sent Abdullah. It was a weird replay of the 1953 events, once again enacted in Karan Mahal. This time Abdullah was free and his arch opponent Bakshi was in detention.

The reaction to the arrest was mild. Sixteen of Bakshi's supporters wrote to me protesting against it and urging that the Assembly be reconvened at once, and fourteen of them went to Delhi to meet the Prime Minister and plead Bakshi's case. Abdullah naturally tried hard to exploit the situation to his political advantage by reiterating that the arrest was 'a very bad practice for democratic functioning', and highlighted the 'non-representative' character of the Assembly. In his speeches at Hazratbal and the shrine of Char-i-Sharif he continued to attack the arrest and indirectly praised Bakshi. This fitted in with an analysis that directly or indirectly he had blessed Bakshi's moves to destablize the State government so as to pave the way for fresh elections.

I made several suggestions to the Government of India in my report of 30 September, including an immediate expansion of the Cabinet, reconvening of the Assembly before the end of October, gearing up the enfeebled State administration and pressing on with the inquiry into the theft of the holy relic. In effect I found myself drawn increasingly into the political vortex. Far from being a detached and aloof constitutional head, I was rapidly becoming a major actor on the political stage. As it happened, I was by that time getting progressively more interested in politics and was beginning to strain at the leash which the headship of the State inevitably constituted. In October, events at last began to move on the lines I had suggested. The Legislature was reconvened and had a five-day session in which, although a formal vote of confidence was not sought, the smooth passage of the supplementary demands for grants proved quite clearly that the Sadiq government was secure, at least for the time being. Sadiq made a long statement on recent political developments which resulted in Bakshi's arrest, and made the point that he had to be detained because, in his desperation to prevent a full-scale inquiry into his past deeds, he had begun acting in a reckless fashion that would have endangered the stability and security of the State. The Ministry was also expanded with the addition to the Cabinet of G. L. Dogra, Mohammad Ayub Khan and Pir Gyasuddin, while Trilochan Dutta joined Mir Qasim as a General Secretary of the party.

By the time we moved down again to Jammu for the winter, the political situation had developed so that the Plebiscite Front came to dominate the scene in the valley. At its convention held at Sopore in the middle of November, Abdullah and Beg took an aggressive stand that clearly indicated their unwillingness to accept the existing constitutional relationship between the State and the rest of India. While they did not mention the word 'plebiscite', the very name of the party made their basic position regarding 'self-determination' clear. At the Governors' Conference later in the month, the first held since Jawaharlal Nehru's passing away, the Prime Minister, Lal Bahadur Shastri, made special mention of the Kashmir situation in his opening remarks. He came out strongly in support of the Sadiq government and said that, while it would be wrong to minimize Sheikh Abdullah's strength, the Government of India was keeping an eye on Kashmir and would take appropriate measures whenever considered necessary.

Lal Bahadur provided a marked contrast to Jawaharlal in many respects. Where Jawaharlal spoke from a high intellectual plane and a broad historical perspective, Lal Bahadur was more down to earth and practical. Where Jawaharlal's stance was that of an unchallenged elder statesman and leader, Lal Bahadur always took a modest and low-key attitude. While I greatly admired Jawaharlal, I always found it somewhat difficult to speak to him on equal terms. With Lal Bahadur I had no such problem. I had come to know him quite well over the previous years, and he had several times called on me at our Delhi residence. I always spoke to him openly and without reservation, which I think he appreciated. At the 1964 Governors' Conference, apart from talking about Kashmir, I also stressed the need to give priority to population control and suggested the setting up of a high-power commission for the purpose.

After the Conference I again went on my annual visit to Mirtola, which I had undertaken ever since 1958. Over these years, my regular meetings with Gopalda and Ashishda and my extensive correspondence with them had become a major factor in my inner growth. I gained not only advice on specific problems but a deeper world-view, a more mature approach to life and its complexities. It is difficult to describe the sort of influence they had on me because it was essentially intangible, but clearly they gave me a dimension which I would otherwise have missed, an inspiration which I would otherwise have been denied. This time Beryl Stileman was out from England to spend the winter with us, so Asha and Beryl decided to come along with me to Mirtola for the first time. Gopalda was delighted to see them, and both he and Ashishda took an immediate liking to Asha.

We spent three nights there, and had long conversations. By then the major question before me was whether or not to get out of the State and plunge into national politics. When I had mooted this in 1958 they had reacted negatively and advised against it, but now six years later they seemed to feel that it might in fact be a positive step, and that if an inner urge to enter a wider sphere had arisen in me it should not be suppressed. This 'clearance' was very important for me, and paved the way for the decisions that I took over the next couple of years. On the morning of 2 December, exactly twenty years after Gopalda's Guru Sri Yashoda Mai had passed away, I got up early and walked up the hill to the Post Office at the top and beyond to greet the sunrise. It was fresh and virginal, and as I walked up the incline the sun came up behind a tree. For a breathtaking moment the tree obscured the sun's disc, sending its rays shooting gloriously in all directions. I walked on spellbound towards the tree, and sat under it reciting aloud in Sanskrit from the second chapter of the *Mundaka Upanishad*:

> In the supreme golden sheath
> lies the Brahman; stainless, without divisions.

A splendour it is; the Light of all lights.
It is That which the knowers of the Self know.

After a while I started walking back to the *ashram.* The sun was now up and shining directly behind me, throwing my full shadow clearly on the trunk of a large tree. As I walked towards it I got the curious feeling that I was becoming aware of a new dimension of myself, that I was looking at myself as it were from a place apart. This encounter with the sun in Mirtola led me to design my personal crest, the sun with a trident inside, and to inscribe thereunder a Sanskrit verse from another Upanishad, the *Shwetashwatara*:

I know that Great Being, resplendent like the sun,
shining in splendour beyond the darkness.

Soon after returning from this trip I went to Bombay to deliver a series of three lectures at the Bharatiya Vidya Bhavan on the 'Post-Independence Generation: Challenge and Response'. Dr K. M. Munshi and his wife Lilawati had always been very friendly, and I admired his deep commitment to the regeneration and reintegration of our cultural and spiritual values. The Bharatiya Vidya Bhavan, which he founded and nurtured till the end of his life, was a unique institution, and continues to play a significant role not only in India but also abroad. In the midst of these positive developments I suffered another physical setback. This time it was a kidney that began giving trouble. There is something in the Kashmir water that tends to develop stones in the kidney. Sadiq had for decades been a kidney patient, and in fact one of his kidneys had been removed many years earlier. D. P. Dhar also suffered from the same complaint, and with two of his Ministers so afflicted, could the *Sadar-i-Riyasat* be far behind? The first attack came in Srinagar in the middle of 1964; the pain was excruciating, and whenever I had the attacks the pain could only be relieved by several intravenous injections of pain-killing drugs. Acute physical pain has a nightmarish quality about it, and after all that I had gone through with my broken legs I had erroneously concluded that my quota was full. I tried all sorts of medicines—Ayurvedic, Allopathic, Homoeopathic—but the stone in the left kidney moved only very slowly and very painfully.

At the Convocation of Banaras Hindu University that winter

I had occasion to confer an honorary degree upon the Prime Minister, Lal Bahadur Shastri, who addressed the Convocation. By now I was becoming closely involved in national affairs. In Delhi I would meet President Radhakrishnan and several Cabinet Ministers. Clovis Maksoud's home was the venue for many stimulating discussions with Indian and foreign leaders, while Romesh and Raj Thapar introduced us to some of their talented friends in journalism and related fields. Looking back at this period I can see that I was rapidly outgrowing State politics and had begun seeking a suitable opening to enter a broader sphere of national activity. It was in this context that on 1 March 1965 I wrote to Lal Bahadur as follows:

I am writing this regarding a personal matter in which I would seek your advice and guidance. I had meant to discuss it with you when I was last in Delhi, but you were so preoccupied with the situation in South India and other pressing problems that I did not consider myself justified in imposing further upon your time.

As you know, I have been Head of the State for almost sixteen years, having taken up my first appointment as Regent in June 1949 at the age of eighteen. I was elected Sadar-i-Riyasat in November 1952, and am now in my third term. During these years I have sought to shoulder the responsibilities and fulfil the duties that devolved upon me to the best of my ability. In a few days I will be thirty-four, and for some time I have been feeling that it would be desirable for me to widen my field of experience. However, the national interest being necessarily uppermost in my mind, I realized that until now my quitting the present post might have created avoidable complications. With the forthcoming amendment of the State Constitution the Office of Sadar-i-Riyasat elected by the State Assembly will be replaced by a Governor appointed directly by the President, and in view of this development I do not think my personal presence here remains as important as it was previously. It is not desirable for anyone to remain too long in the same public office, and a change would now seem to be advantageous. With the advent of the office of Governor a wide choice of talent is now available, and I could offer my suggestions regarding a suitable person for this important appointment.

With regard to myself, I am seriously considering contesting the next general elections, specially as the people of this State will for the first time be electing their representatives to the Lok Sabha directly. Also, there is a general feeling that younger people should take a more active interest in politics, and it is possible that with such experience as I may have gained by being in public life since an early age I could if elected be of some service to the nation on a broader canvas.

The next elections are exactly two years away, and it is with special regard to the intervening period that I am writing this letter. I understand that the present High Commissioner in the United Kingdom Dr Jivraj Mehta is likely to be replaced soon. It is embarrassing for me to write this, but I do feel that perhaps I could fill that appointment successfully. At this crucial juncture in world affairs our diplomatic posture abroad is of the utmost importance, and England still has special significance from our point of view. I have been there several times, and such acquaintance as I have with English life, language and politics should be a valuable asset. It would also provide me with a new dimension of experience before I enter the mainstream of national politics.

Another personal aspect of considerable importance to me is the fact that my mother has been ailing for the last several years, and despite the best medical advice in Bombay and Delhi she has shown no sign of improvement. Her condition is causing me grave anxiety, and I feel that if I can take her to London with me for a prolonged course of treatment she might improve. The Chaubarsi ceremony of my late father falls on 11 May this year, after which my mother will be able to undertake a foreign trip.

It is with reluctance that I am inflicting upon you a letter concerning my own personal future, but knowing as I do your affectionate and understanding temperament I have written quite frankly and look forward to getting the benefit of your guidance and encouragement at your earliest convenience.

The Prime Minister replied immediately saying that he would discuss the matter with me when we next met. This he did in April, when he said that in view of the uncertain position in the State he would not like me to be away at the present juncture. As it happened, his Principal Secretary, L. K. Jha, called on me the next year offering me the assignment, but by then certain political developments had taken place which impelled me to decline it. If the offer had been made when I was interested, the whole of my future life would have been in diplomacy rather than politics. On such tenuous circumstances does our destiny hinge.

Early in 1965 Articles 356 and 357 of the Indian Constitution were extended to the State. These contained provisions in the event of failure of constitutional machinery in the States, and the step caused an immediate outcry from Sheikh Abdullah at his speech in Hazratbal the following Friday. He launched an inflammatory diatribe against India, both for extending these articles and also for the move to merge the National Conference with the Indian National Congress. Bringing up his pet bogey, he said that this would mean the end of the Muslims in the State and that they should be prepared to offer every sacrifice to safeguard their interests. He also advocated a social boycott of National Conference MLAs who had joined the Congress, and dismissed Lal Bahadur Shastri as a weak man from whom nothing could be expected. This speech caused a serious debate in New Delhi as to whether Sheikh Abdullah should be allowed to fulfil his long declared intention of going abroad on Haj and to other countries, particularly as we received intelligence reports that he had told his followers that he would contact leaders in foreign countries to muster support for self-determination in Kashmir. It was finally decided that he should be allowed to go, accompanied by his *alter ego* Afzal Beg, and that no restrictions should be placed on his movements abroad. Lal Bahadur shrewdly felt that if he misused the freedom he would stand self-exposed, while if travel curbs were put on him he might refuse to go and make a big political issue out of it.

On 26 January 1965 the Indian National Congress was formally launched in the State, thus marking the beginning of a new chapter in its chequered history and fulfilling a goal for which I had personally laboured hard from behind the scenes for some years. I believed that the entry of the premier national political party in the State would be important in strengthening the integration of Jammu and Kashmir with the rest of the country. While constitutional, legal, financial and administrative ties were important, in a democratic system it was political parties that provided the flesh and blood to the functioning of the whole system. And, quite clearly, the Congress, with its magnificent record of national service first under giants like Sri Aurobindo and Lokmanya Tilak and later under Mahatma Gandhi and his brilliant band led by Jawaharlal Nehru, was clearly the predominant political force in India and should certainly extend its activities to the State.

Soon thereafter, the government announced the decision to appoint a Commission of Inquiry consisting of Justice Rajagopala Ayyangar to inquire into thirty-eight specific allegations of corruption and misuse of power by Bakshi Ghulam Mohammad. While under detention at Udhampur, Bakshi had suddenly developed symptoms of cardiac arrest. Doctors had been rushed to treat him, and he had subsequently been released from detention. A third important event that month, with which I was also closely involved, was the opening of the new social sciences block of the University campus at Jammu by the Union Education Minister M. C. Chagla. I had for some time been moving towards a bifurcation of Jammu and Kashmir University, and this was a major step in that direction.

The anti-Hindi riots in Madras also erupted that month, in which many lives were lost. I wrote to the President on 1 February:

> The recent anti-Hindi demonstrations in the South, with the horrifying violence and incidents of self-immolation that accompanied them, have been a source of acute pain and distress to all right-minded people. Without making any detailed comments on this whole issue, I would like to draw a parallel between the secessionist ideology of the DMK in the extreme South and that of the Plebiscite Front in the extreme North. The former has already led to tragedy, and the latter has equally grim potentialities. An additional danger here is the communal factor,. with Pakistan constantly fanning the flames of religious hatred and a powerful leader like Sheikh Abdullah playing an openly secessionist role. In fact there are at present all the elements that go to create a really explosive situation, and we must be fully aware of them.

In February, after spending some days in Delhi, Abdullah left for Bombay and from there took a boat for Haj. He made no secret of the fact that he was going to use his pilgrimage for political purposes. Indeed this blatant exploitation of religion for political purposes was one of his special techniques ever since he emerged into prominence in 1931. Despite his much-professed secularism, it was almost invariably on a religious platform that he enunciated his political policies. Before he left he had succeeded in creating a good deal of anti-Indian sentiments in the valley, and the social boycott was beginning to have a disturbing effect on pro-Indian elements among the

Muslims. There were cases in which dead relations of Congress workers were refused burial space, and many other cases of harassment and discrimination began taking place. I referred indirectly to this in my address to the joint session of the Legislature in Jammu on 3 March in which I said that 'there have been indications of types of political and social activity which tend to spread the poison of communal exclusiveness, and sometimes even hatred, and which may gravely jeopardize the peaceful development of democracy. Government is aware of this problem, and does not propose to permit the creation of communal or separatist tendencies to be carried to dangerous limits of poisoning or inflaming public opinion.'

In the first week of the session the government decided to move on two points to counter the activities of Sheikh Abdullah's party. On the night of 6/7 March large-scale arrests were made of Plebiscite Front leaders and workers in the valley. It was a surprise operation, and about two hundred people were rounded up without any force having to be used. Simultaneously a Constitution Amendment Bill was introduced in the Assembly providing for a Governor to be appointed by the President in place of the *Sadar-i-Riyasat* elected by the State Assembly, and also for change in the nomenclature of the head of the government from Prime Minister to Chief Minister. These were symbolic measures designed to bring the State more closely on a par with other constituents of the republic, but in public life symbols can be of deep political significance. I was, of course, delighted, because I looked upon the abolition of my office as being my major achievement. The amendment bill was passed and I accorded my assent to it on 10 April. That very evening the Chief Justice administered to me the oath of office as Governor after the appointment letter was received from the President. Thus I became the first and last *Sadar-i-Riyasat* in history, having held that office for just over twelve years. My first public function as Governor was to host a reception on 13 April on which, by a happy chance, both Baisakhi and Id-ul-Zuha coincided that year.

In the valley the Plebiscite Front launched a series of protest meetings against these measures at which strongly anti-Indian speeches were made, while abroad Sheikh Abdullah, as I had expected, began acting in a provocative manner. He met

Premier Chou En-lai and accepted an invitation to visit that country. There was an outcry in the press and in Parliament, and finally the Minister for External Affairs announced that all endorsements on the passports of Abdullah and his party had been cancelled except those necessary for the purpose of Haj. This gave rise to acute speculation as to what his further moves would be. Would he return to face the consequences of his actions or would he continue to operate from abroad? Knowing him, I felt sure he would return. Whether one agreed with his politics or not, there was no doubt that he was a man of courage and determination. It was not for nothing that he had come to be known as the Lion of Kashmir.

In the event, Sheikh Abdullah, the Begum and Mirza Afzal Beg returned to India on 8 May, were taken into custody by the Government of India and flown to Ootacamund in south India where they were ordered to remain within municipal limits. So, for the third time since 1947, Abdullah found himself under detention. His two main failings were his unwillingness to accept that Jammu and Kashmir would have to continue as a part of India like other States, even though it may enjoy a somewhat special status in some matters, and also his inability to come to grips with the tri-regional character of the State in which the aspirations of the people of Jammu and Ladakh were often contradictory to those of the valley. Indeed, this question of regional aspirations was emerging as a major factor in State politics, and I became actively involved soon thereafter in a full-scale controversy connected with this, which, in turn, paved the way for my entry into the political arena.

Chapter 23

In April 1965 Asha and I went on a fortnight's holiday to Japan. I was deeply impressed by the tremendous dynamism and vitality of the Japanese people, combining all the benefits of technological advance while retaining a distinctive quality reflecting their cultural traditions. They seemed to combine an infinite capacity for hard and disciplined work with an artistic flair that converts an ordinary park into a work of art. The Buddhist temples—the golden and the silver pavilions—and the Ryoan-ji temple at Kyoto inspired by Zen Masters were particularly fascinating. A beautiful fifteenth-century shrine, the Ryoan-ji temple consists of a rectangle of pure white sand carefully raked in broad parallel lines, with fifteen rocks of varying sizes thereon. Through the centuries observers have read all sorts of symbolic meanings into this garden. I was immediately reminded of photographs taken of interstellar galaxies. Each group of rocks seemed to symbolize these mighty galaxies whirling through space, creating their own universes of time and consciousness.

Unfortunately my kidney attacks had become increasingly frequent, and I had a dreadful bout at Kyoto Hotel from where Asha and our devoted Secretary, Milap Chand, rushed me to a hospital. Renal colic can be one of the most painful of conditions, and my stone was determined to give me as much trouble as it could. Despite this, the Japanese visit was a memorable one. The Shinto shrines, with their stark simplicity, were as impressive as the Buddhist ones, especially the celebrated Meiji shrine which is an oasis of peace in the heart of bustling Tokyo. It was while we were in Japan that the skirmish with Pakistan in the Rann of Kutch took place, which turned out to be a sort of curtain-raiser for a more serious conflict.

Ever since the breakdown of direct negotiations with Pakistan, activity on the Ceasefire Line had increased, and President Ayub began making increasingly bellicose statements. In retrospect, it is obvious that he was in fact preparing the ground for a last attempt to grab Kashmir, the obsession that led so many of the Pakistani leaders to their doom. On 4 June I said in my letter to the President that 'it is clear that we are now closer to war with that country than we have ever been since 1947'. The war earlier that year in the Rann of Kutch, although ending in a stalemate, was a sort of probing operation on the part of the Pakistanis to test India's mettle under Lal Bahadur, now that the towering figure of Jawaharlal Nehru had been removed by death.

In the midst of all this growing tension my mother suffered a severe cerebral stroke. She had kept very indifferent health since my father had passed away, and a combination of long-standing diabetes and prolonged austerity in the year after his death had caused severe medical problems. She had moved into the new wing that I had built for her at Karan Mahal, and it was there that she suffered the stroke. For some days we were not sure that she would survive, but she managed to pull through, although the attack left her partially paralysed and with impaired speech. It was heart-rending for me to see this once beautiful and vivacious woman, so fond of company and jewellery and colourful clothes, reduced to an invalid unable to leave her bed. We did all we could to comfort her, and as she loved to hear me sing I would render devotional songs every evening in her presence.

The Northern Zonal Council met in Srinagar early in July. It always seems to me to have been a pity that this concept of Zonal Councils made by the States' Reorganization Commission never really got off the ground. Meetings have been held perfunctorily from time to time, but the Councils have never really made any meaningful contribution to the integration and development of the States concerned. This one was presided over by the Home Minister, Gulzarilal Nanda. The Action Committee had meanwhile launched a civil disobedience movement, and Maulana Farouq's Awami Action Committee—clearly drawing its inspiration from Pakistan—adopted a constitution pledging to secure for the people of the State 'the right of self-determination'. This movement continued for several weeks, when suddenly a series of devastating fires broke out in different parts of the valley. It was a measure of our complacency and a signal failure of intelligence that even in the beginning of August we did not have any information regarding the diabolic move that Pakistan had planned to overrun Kashmir. In fact, on 5 August I gave a large reception at the Oberoi Palace Hotel to celebrate Vikram's first birthday, and it was in the course of that party that I first heard whispers that a few infiltrators had entered the valley. I did not realize how serious the situation would become, and so on the 6th I went to Delhi for a Sanskrit conference at the Prime Minister's house.

Since early that year Gopalda had been quite ill in Mirtola, and by June he had to be brought to hospital in Naini Tal. I was deeply disturbed about this, and decided to make a quick visit to see him. Nasib was with me; he was a good friend and travelling companion, and we could chat at length during the trip. When we reached Naini Tal, where Gopalda and Ashishda were staying with one of their disciples, Jagdish Shah, I was shocked to see Gopalda. He looked dreadfully ill and had changed almost beyond recognition from the robust, vigorous man I had met a few years earlier. He was suffering from intestinal obstruction and was in great pain, but he greeted me with his usual warm affection and retained his spiritual presence despite his serious illness. We spent the night at a hotel and took leave of him the next morning. When I bowed to touch his feet he embraced me warmly and held me to him for several moments. His flashing blue eyes were filled with love and power; I think we both knew then that we would never see each other again in this life. Of all the men I have ever met, he was the greatest. Of him do the immortal words of the Upanishad ring true: 'Sporting in the Atman, absorbed in the Atman, constantly in action, they are the foremost among the knowers of Brâhman.'

By the time I returned to Delhi the news from Kashmir had broken. Incredibly enough, despite the bulk of the Indian army being spread over large parts of the State, and an administration supposedly in control of the situation, thousands of well-trained, highly motivated and well-equipped guerillas from Pakistan had been able to enter the valley and some had virtually reached the outskirts of Srinagar. Evidently the result of long and careful planning, this was Pakistan's last major attempt to get by might what it had always believed was its by right. It must be admitted that India's intelligence failure was as serious as that which occurred during the Chinese debacle. I hurried back to Srinagar, to find Sadiq and his colleagues quite shaken by the turn of events. Even the normally unflappable D. P. Dhar had lost some of his poise, because as Home Minister it was clearly his responsibility to have been aware of this possible threat. To this day I find the triple breakdown of army, Central and State Intelligence to be astounding.

A guerilla invasion of this scale must have required many months, if not years, of training and planning in Pakistan and Pakistan-occupied Kashmir, but we seemed to be blissfully ignorant of the whole matter. The invasion was accompanied by a virulent and carefully orchestrated propaganda barrage. Huge posters entitled 'Proclamation of War of Liberation' appeared in English and Urdu in various parts of the valley, issued by 'the Revolutionary Council of Kashmir'. They claimed to have been published in the 'Sada-e-Kashmir Press, Srinagar', but no such press existed and obviously they had been smuggled in from Pakistan. Clearly, Field Marshal Ayub banked on a massive uprising by the people of Kashmir against 'Indian Imperialism'. The fact that no such uprising took place, and that the people of the valley remained loyal to India, must have come as a rude shock to the military rulers of Pakistan.

The invasion threw the State administration into disarray. I joined Sadiq and his colleagues in helping to restore some order and balance. Several meetings were held at my residence. We were particularly worried about Srinagar city, which had always harboured a hard core of pro-Pakistan elements. If the city or its weaker points were to fall even temporarily under the control of the guerillas, it would be a major disaster. We, therefore, concentrated especially on Srinagar as well as the three other major towns in the valley—Anantnag, Baramulla and Sopore. There were greatly varying estimates regarding the number of guerillas involved, and it has never been possible to get an accurate figure. My own guess is that about five thousand people were directly involved, apart from the massive backup from bases across the Ceasefire Line. Their pattern of operation was of the typically guerilla type; they would avoid direct clashes with our security forces and attack at points and timings chosen by them.

The invasion was by no means confined to the valley. In Jammu there were major incursions in the Poonch–Rajouri sector, and infiltration took place deep into the Gool–Gulabgarh area of Udhampur district. Heavy artillery attacks were launched on the Ceasefire Line in Jammu and also across the international frontier. The luckless refugees who had settled in the border area of Chhamb–Jaurian to the west of the Chenab became restive and started to stream into Jammu city. It was a tribute to the people of the State that despite this cruel provocation there were no communal riots and the Hindu–Muslim relationship remained cordial.

On 25 August the Indian army crossed the Ceasefire Line in the Kargil sector, reoccupied two posts opposite Kargil that had earlier been vacated, and also launched attacks in the Tithwal and Uri sectors. As far back as 1951 Jawaharlal Nehru had warned Liaquat Ali Khan that any Pakistani attack on Jammu and Kashmir would be construed as an attack on India and would cause the area of war to be extended. Evidently, this lesson was lost on President Ayub Khan or, more likely, he thought that Lal Bahadur would simply not have the courage to hit back at Pakistan. The whole tenor of the Pakistani press ever since its inception was to pour scorn on the 'meek and servile Hindu', the 'dhotiwalas' who had been ruled for centuries by the Muslims. It seems that the rulers of Pakistan became victims of their own propaganda. Their first major miscalculation was the Kashmir operation in 1965; their second, seven years later, resulted in the defeat and breakup of their country itself.

Our army moved effectively into Pakistan-occupied Kashmir, the most important operation being the capture of the bulge between Uri and Poonch which involved overrunning about two dozen Pakistani posts. This gave us a much stronger strategic position and made it more difficult for the infiltrators to enter both the valley and the Jammu region. Soon thereafter Pakistan launched a major armoured attack in our vulnerable Chhamb–Jaurian sector in Jammu, west of Manawar Tawi, and we responded by launching a major offensive on the Jammu–Sialkot border. Thus, what started as guerilla infiltration developed into the first full-scale war between India and Pakistan since the Jammu and Kashmir operations in 1947–8. Lal Bahadur Shastri had clearly reiterated Jawaharlal's position that an attack on Jammu and Kashmir would be treated as a full-scale invasion of India, but President Ayub evidently still believed that Jammu and Kashmir being a 'disputed State', the war would remain confined to that sector only.

Once the war began, it developed into a full-fledged collision. The major battles were fought in the plains of Punjab on both sides, because that was where heavy armour, artillery and air strike power could be fully deployed. Our forces made initial progress but were stopped just short of Sialkot and the outskirts of Lahore. Had either of those towns fallen, the result of the war would decisively have been in India's favour. However, the whole power structure of Pakistan, then as now, revolved around Punjabi Muslims who also constitute the bulk of their armed forces. With our attack on their heartland, therefore, they had to abandon the plan to capture Jammu and Kashmir and instead were forced to regroup for defence. On our part, the boastful claim that Pakistani tanks would break through our defences and have a 'cake-walk' to Delhi was thwarted by the gallant defence put up by Indian troops in the Punjab sector. Pakistan was able to overrun the basically indefensible strip in Jammu to the west of the Manawar Tawi, but was not able to get anywhere near Jammu town.

I donned my Major-General's uniform and visited the forward areas, including those where our tanks were operating well within Pakistan territory under the command of General Rajinder Singh 'Sparrow', in what were the largest tank battles since the end of the Second World War. In Jammu I had my first experience of being under an air raid. Jammu is hardly three minutes flying time away from Pakistan, and their air force made several attacks on the vital Tawi Bridge—which they were unable to damage—and at various points along the Pathankot–Jammu highway, which was a crucial lifeline for defence and communication purposes. I had reached Jammu in the afternoon along with Nasib, and was in Hari Niwas when the air-raid sirens began wailing. Simultaneously, I had another one of those wretched kidney attacks which left me writhing in pain. The doctor who was called to give me an intravenous injection was so nervous that he dropped the syringe twice, and we had to have it sterilized repeatedly. It was a nasty experience, but luckily both the kidney pain and the air attacks passed off without any major casualty.

Apart from the danger of air attack, the major problem in Jammu was the huge influx of refugees from the border tehsils of Akhnoor and Ranbirsinghpura, which involved about two hundred thousand people. In the Poonch–Rajouri area the Hindu population started converging upon the towns for safety. Thus, these unfortunate refugees were for the second time in two decades uprooted from their hearths and homes. It is a cruel irony that, despite hundreds of crores of rupees spent in the State, the refugee problem has still not been satisfactorily solved. Mahavir Tyagi was then Rehabilitation Minister, a jocular but not very substantial political figure, and though he tried to be helpful he was not really able to get a grip on the problem. We sent all the Jammu Ministers of the Sadiq government down to various parts of that region, and while there I set up a Citizens Council which included all shades of public opinion as well as key administrative officers. By and large, the Congress and the Jana Sangh overcame their political differences in the face of the grave threat to national security. In the valley, however, it was a different story. Maulvi Farouq openly espoused the cause of the raiders in his religious sermons, and even the wily Maulana Masoodi came out with a statement that a plebiscite was the only solution to the Kashmir problem, and demanded the release of Sheikh Abdullah and other political detenus. While Abdullah himself was silent, one of his younger sons, who was living in England, made a statement strongly attacking India and supporting the 'freedom fighters' who were attempting to 'liberate Kashmir'.

A curious feature of the national scene was that, even before the war began, Sant Fateh Singh had started an agitation for a Punjabi Suba and had threatened to go on a 'fast unto death' to achieve this. Among others, Sadiq appealed to him not to take this step in view of the situation that had arisen in Kashmir, and finally as a result of hectic political activity and the intervention of such respected public figures as Sardar Hukum Singh, Speaker of the Lok Sabha, the Sant decided at the last minute to postpone his fast. My comments on this in my letter dated 15 September 1965 to Dr Radhakrishnan bear quoting, because even today, twenty years later, the problem remains a major source of tension and a grave threat to national security and integration. I wrote:

Sant Fateh Singh's last-minute decision to postpone his fast has indeed been extremely welcome, as nothing would have been more disastrous at this stage than any Hindu–Sikh tension in the Punjab. I would submit, however, that the problem has really not been solved but only postponed. Surely, as long as there is no question whatsoever of secession or any possibility thereof, it should not be beyond our ingenuity and good sense to evolve a formula whereby the brave Sikh community feels politically and psychologically satisfied. Indeed this must receive top priority, because the continued failure to solve this problem can only be extremely detrimental to our national interests, specially in the context of our confrontation with Pakistan.

The war with Pakistan immediately evoked a tremendous upsurge of patriotic fervour throughout the country. From north to south, from east to west, India rose as one man to confront the aggressor. This heart-warming phenomenon belied the gloomy predictions of those foreign 'experts' who had expected India to break up at the first major crisis. I went to Delhi to meet Prime Minister Shastri and found that this humble and diminutive man had suddenly risen in stature in the eyes of his countrymen. His National Relief Fund, to which I made a personal contribution of a hundred thousand rupees, evoked tremendous support. Voluntary organizations in all parts of the country contributed to the collection drive. Most impressive was the response of the brave citizens of Punjab, Haryana and Jammu who prepared food for the troops in the forward areas and opened free canteens on all the major routes along which our armed forces would pass. Women in particular helped to feed and nurse the injured jawans in the hospitals to which they were evacuated from the battlefield.

As a result of the intervention of the two super-powers and the United Nations, a ceasefire was announced by both sides at the end of September. At that time we were in possession of the vital Haji Pir bulge between Poonch and Uri, and also some territory in the Sialkot sector. We had lost the Chhamb segment in Jammu, and the Pakistanis had succeeded in making inroads in the Punjab sector. The debate at the United Nations followed predictable lines, with the talented but fanatically anti-Indian Zulfikar Ali Bhutto leading the Pakistan side, while our delegation was led by the cool and unflappable Sardar Swaran Singh. Mir Qasim was a member of the Indian delegation, and made some excellent speeches rebutting Bhutto's blatantly communal approach to the whole dispute. Finally, the Security Council adopted a resolution saying that subsequent to the ceasefire there should be a 'withdrawal of all armed personnel back to the positions held by them before August 5th 1965'.

By this time, their invasion having failed, the raiders began slipping back across the Ceasefire Line. In the same way as we were unable to prevent their entry, we failed to prevent their escape, and this showed up a major weakness in our strategic situation. In the valley, anti-Indian elements began surfacing in various educational institutions, especially the Regional Engineering College near Hazratbal, which remained for several years a hotbed of anti-national activities. Ultimately, in view of their inflammatory speeches, Moulvi Farouq and four other anti-India leaders were arrested on 10 October and taken down to Jammu. Simultaneously, curfew was imposed in some parts of the city for three days, during which the police had to open fire several times on unruly mobs, unfortunately resulting in about half a dozen deaths. On 20 October Maulana Masoodi, Ghulam Mohiuddin Quarra and several others were also arrested and sent down to Jammu.

Following these arrests there was a spurt in terrorist activities such as incendiarism and bomb throwing. Provocative and highly inflammatory posters, hand-written and cyclostyled, began appearing in various parts of Srinagar using the most abusive anti-Indian and anti-Hindu language and giving detailed instructions regarding terrorist activities, urging people to cut telegraph and telephone wires, to set fire to Indian vehicles and kill 'Bharati beasts' wherever possible. Clearly this was part of the activity planned by Pakistan for invasion, and was carried out by trained agents who stayed on in the valley after the bulk of the guerillas had withdrawn. Our security forces took necessary measures to counter these activities, and we posted special guards at the Hazratbal mosque as we had received reports that in their desperation the Pakistani agents may try and blow up the sacred shrine itself in order to pin the sacrilege on India and further inflame Muslim sentiments.

During the entire period I remained very active, especially in touring the refugee camps in Jammu and trying to see what could be done to improve their lot. Asha mobilized the Ladies Club at Jammu and Srinagar to undertake a whole range of welfare activities for the refugees, particularly women and children. Almost two decades earlier my parents, especially my mother, had been involved in the same sort of activities. This time she was incapacitated by her illness, but Asha stepped into the breach. In fact, it was because of my mother that during the war we moved out of Karan Mahal, which was a prime target for an air raid, and shifted into the orchard house called Cherry Grove. Apart from visiting refugee camps, I toured the main towns of the Jammu region, exhorting people to remain vigilant and united, and also paid several visits to our troops in the forward areas. In the capital I addressed Delhi University students and also spoke at other forums, addressing myself to a wider national audience.

I was also taking an active interest in the State politics. A glaring feature of the situation was that ever since the advent to power of the National Conference in 1949, political power in the State had come to be vested almost exclusively in Kashmiri politicians, and Jammu was relegated virtually to a subsidiary position. Dogra rule had in effect been replaced by Kashmiri rule, and the unfair imbalance between Jammu and Kashmir that had existed ever since my ancestors founded the State in the middle of the nineteenth century continued in reverse form. In Sheikh Abdullah's Cabinet there had been only one Cabinet Minister from Jammu, and though the position improved marginally under subsequent regimes, the fact remained that effective political and economic power continued to vest in Kashmir. When the State Constitution had been framed there had been a realization of this problem, and a half-hearted attempt to meet it had been made by providing for parity between the two regions in the 36-member Legislative Council, but this was never followed to its logical conclusion and Jammu remained very much of a political colony of Kashmir leadership.

This aspect had been weighing on my mind for many years.

In a way, my appointment first as *Sadar-i-Riyasat* and then as Governor was tacit recognition of the Dogra factor in State politics, and with my mind now set on entering national politics I was looking around for a way to ensure for Jammu a place of honour and dignity. Clearly, this could be done only in one of the two ways. Either the logic of the States' Reorganization Commission, whose jurisdiction had specifically been excluded as far as Jammu and Kashmir was concerned, should now be made applicable to the State. This would obviously result, as it did in the case of the erstwhile Nizam's State of Hyderabad, in a trifurcation of the State. Alternatively, the State Constitution could be substantially amended to ensure full political and economic participation for both the major regions, as well as special attention to the far-flung backward area of Ladakh.

I was in Delhi again in the middle of November for the Governors' Conference. I had been discussing this whole matter with some close friends, and Romesh Thapar wrote up my ideas in his column in the *Economic and Political Weekly*. On 14 November, which happened to be Jawaharlal Nehru's birth anniversary, I received a telegram from Ashishda saying that Sri Krishnaprem had passed away that morning in Naini Tal. I was upset and disturbed. That evening I happened to go to Ashoka Hotel where I ran into Neville Maxwell, then the correspondent of the *London Times*. He asked me to elucidate my views on the future of the State in the light of the comments in Romesh's article. I did so in somewhat forthright terms, and he sent the despatch back to his newspaper.

When this was published in London it was immediately flashed to India and hit the headlines. Meanwhile, Kuldip Nayar, then heading the UNI, had an interview with me which he released a couple of days later. I reproduce the text of his despatch:

> The Sadar-i-Riyasat of Jammu and Kashmir Mr Karan Singh is of the opinion that there should be a unilingual Kashmiri-speaking State.
>
> Making it clear that he was speaking only in his individual capacity, Mr Karan Singh told a UNI correspondent that Jammu be merged with Himachal Pradesh, and Ladakh should become a centrally administered area.

This arrangement, Mr Karan Singh said, would leave the area beyond the Banihal Pass as a separate unit, facilitating the formation of a new state within the Indian union.

Mr Karan Singh said that at the time of reorganization of states, the State of Jammu and Kashmir was excluded from the S.R.C.'s purview on the plea that the issue relating to the region was pending in the United Nations. Now that a cabinet sub-committee was re-examining the question of the formation of a Punjabi Suba, the reorganization of Jammu and Kashmir could also be entrusted to the same committee.

Mr Karan Singh said that he would not like to say anything on the formation of a Punjabi-speaking state. But if the Punjab was to be organized and if hill districts were to be excluded, then those districts should be merged with Jammu and Himachal Pradesh so as to form a viable hill state.

Mr Karan Singh said that the present State of Jammu and Kashmir was an 'administrative monstrosity', with the government moving back and forth between Jammu and Srinagar. This could not continue in present circumstances.

Mr Karan Singh said that there was no sanctity behind the J & K State as it was. In fact, his family had brought the two parts together through conquest, and he as a successor would say that the sooner the present arrangement was ended, the better it would be.

Mr Karan Singh said that the people of Jammu would welcome the partition of the State, and even the people living in the valley would like to have their own separate unit because this would give them their unilingual State, more compact and convenient.

Asked if he had broached this suggestion with leaders in Delhi, Mr Karan Singh said he had already broken the ice.

As soon as this was published all hell broke loose. I had made my comments in good faith, and quite frankly had not realized that it would cause such a commotion. I suddenly found myself under severe attack from almost the entire political spectrum, from the communists on the left, who seemed to think it was part of some sinister American plot, to the Jana Sangh on the right who thought that by isolating Kashmir the Muslims would be so preponderant that we would lose it to Pakistan. There was a furious controversy in the press and Parliament, and I was taken aback at this turn of events. However, there was no question of my resorting to the timeworn politicians' trick of denying the whole story. After careful consideration, I decided that in all fairness I should offer to resign as Governor and squarely face the political consequences. I drafted a statement and took it to Lal Bahadur. It ran as follows:

As there has been some public comment arising from a chance interview that I had recently with a foreign correspondent, I feel it is desirable to clarify the position. First I would emphasize, as I did in the interview itself, that the views I expressed were strictly in my private capacity as a citizen of a free society, and should not be taken to reflect the official attitude of the Jammu and Kashmir Government or the Government of India. Secondly, the basic premise underlying my remarks was that Jammu and Kashmir State is no longer negotiable and must always remain part of India. We have recently fought a war to confirm this national determination, and in this context the possibility of a future reorganization of the State on a regional basis came up for discussion strictly as an internal measure which demands consideration because Jammu and Kashmir had been excluded from States' Reorganization a few years ago. There is scope for honest difference of opinion regarding the desirability or timing of such a move, but in a democracy there is no reason to fear free public discussion or to indulge in the absurd exercise of imputing imaginary motives in the event of genuine disagreement. My view is that once we wholeheartedly accept the fact that Jammu and Kashmir is irrevocably part of the nation, its internal structure—like that of the hundreds of other erstwhile princely States which acceded to India—can no longer be considered permanently unalterable.

Another point I must stress is that the idea of a regional reorganization of Jammu and Kashmir cannot in fairness be dubbed communal merely because the regions concerned happen to have different religious compositions. It is hardly necessary to reiterate that for me secularism and communal harmony are articles of faith. What is important is that the genuine aspirations of Indians living in the three clearly defined and historically demarcated regions of the State—Kashmir, Jammu and Ladakh—must be respected, and their welfare furthered in every way possible so that they can better contribute towards strengthening the foundations of our secular and democratic polity. Facing as we are a grave threat to our freedom and integrity, I would urge that this question be approached in a calm and rational manner, keeping in mind the compulsions of geography and economics, administration and national security.

For over sixteen years—almost half my life—I have had the rare privilege of acting as Head of the Jammu and Kashmir State. Throughout this period my paramount aim has been to further—to the best of my ability—what I saw as the national interest, and to serve the people of the State as far as it lay in my capacity to do so.

It is possible, however, that the controversy raised by my interview—given though it was in perfectly good faith—might have caused some embarrassment to the State Government headed by Shri G. M. Sadiq, for whom I have deep personal regard. Keeping this possibility in view I have ventured to indicate to the Prime Minister that if he so desires I will gladly submit to the President my resignation from the office of Governor of Jammu and Kashmir.

When I handed him the statement I said, 'Lal Bahadur ji, you remember that when you were Railway Minister you resigned over a major accident that occurred. I feel that I should also follow in your footsteps.' He read the paper, then looked up with a mischievous twinkle in his eye. 'So this is your accident', he said, and laughed. The tension was relieved. He advised me not to send the letter but to wait for the storm to blow over. Perhaps he felt that my quitting at that juncture would further complicate matters. But while the controversy regarding my statement gradually subsided, it left a permanent influence on my career. Clearly, I was not going to spend the rest of my life as a ceremonial head, even though as Governor of my own State I did have much wider scope for political activity than others. It was simply a question of how and when I would get out of my job and plunge into national politics. Meanwhile, the development brought me into a direct clash with the man who had for all these years been my closest political confidant—D. P. Dhar.

'D.P.' was a brilliant and patriotic politician, and had rendered valuable service ever since the first Abdullah government in 1949. But he was rather jealous of my special position as the conscience-keeper of the Central Government, and after Jawaharlal's passing I suspect that he began to be a little worried at my growing influence in the Centre. Also, my demand for a fair deal to Jammu seemed to him—quite mistakenly—to be aimed at the Kashmiri Pandit community to which he belonged. In fact he was an agnostic, and in some ways I was a much better 'Pandit' than he. I have always held the highly intelligent Kashmiri Pandit community in high regard, and received almost all my education from them. Nonetheless, it was true that for historical reasons they had a much larger share of the State services than their minuscule population would have warranted. Whatever the reason, D.P.

was in the vanguard of the attack launched on me after my statement. I was surprised and hurt at this, but was not prepared to back down and retaliated in kind. The net result was that our relations became strained, and we never really recovered our cordiality even when we were later colleagues in Indira Gandhi's Cabinet. His death after a heart attack at a comparatively early age removed one of the most elegant and intellectual figures from our public life. I was with him in the Pant Nursing Home a few hours before he died, and although he could only speak with difficulty I held his hand for several minutes, recalling our close friendship and association in the early, turbulent years of our youth.

I had put a paragraph in my letter to the President dated 15 November about Jammu:

> As the result of my tour of the Jammu province I have gathered the impression that the people are suffering from a sense of grievance and neglect. They feel that they were not given adequate attention during the recent conflict, although it was Jammu that bore the brunt of the Pakistani aggression. It would appear that both on the political and the administrative fronts we were not really effective during the crucial days of the conflict. One of the reasons was that the seat of the Government was then in Srinagar, and the provincial administration in Jammu was ill-equipped to cope with such an emergency. Indeed it is quite clear that the old system whereby for six months each Jammu and the valley are deprived of adequate political and administrative leadership is no longer viable. Some method will have to be devised whereby both places are throughout the year given the care and attention that they deserve.

In my next letter, of 8 December, following the raging controversy on my statement, I faced the issue squarely. It was always my policy to place my views clearly and unequivocally on record, whether or not they were palatable at any given point in time. I wrote:

> There has been some controversy and a good deal of public interest in the State as a result of a chance interview that I had with the New Delhi correspondent of the *London Times* in which I supported the reorganization of Jammu and Kashmir on a linguistic basis. I have already called upon you personally in this connection to make a detailed report regarding the incident. In this letter I will not go into any details except to restate the four basic presuppositions—which to

my mind are incontrovertible—upon which the whole idea is based:

(a) Jammu and Kashmir is now no longer negotiable. We have fought a war to confirm this national determination, and therefore any reorganization is purely an internal matter.

(b) Not a single erstwhile princely State today retains its old frontiers, and therefore the argument that there is some special sanctity regarding the boundaries of Jammu and Kashmir is neither logical nor valid.

(c) This State was specifically excluded from the purview of the States' Reorganization Commission, but in the changed circumstances the genuine aspirations of Indians living in the three clearly defined regions of the State must be taken into account.

(d) A reorganization cannot be dubbed communal merely because the regions concerned happen to have different religious compositions.

Continuing my letter I wrote:

I returned to Jammu on the 3rd, and immediately had a very frank discussion with the Chief Minister Shri G. M. Sâdiq and Syed Mir Qasim. Sadiq Sahib met me again this morning before leaving for Delhi. I am happy to report that the excellent relations and mutual understanding that subsisted between us have in no way been impaired despite our genuine and honest disagreement on this matter. A section of the press, however, has been busy trying to play up a 'rift' between me and the Chief Minister, but if neither of us makes any further public statement the controversy is likely to die down. Of course it has generated a good deal of discussion and comment throughout the State, but it would appear that by and large the grassroots reaction in Kashmir, Jammu and Ladakh has been favourable. In any case we have decided that the matter should be allowed to rest where it is for the time being.

Chapter 24

With the United Nations unable to break the deadlock, the Soviet Union entered the picture as a close neighbour both of India and Pakistan and a major power in this part of the globe. It was interesting how in regard to the Indo-Pakistan crisis there almost seemed to be a convergence of interests between the two super-powers. Clearly Pakistan would not have agreed to attend the Tashkent meeting unless President Ayub had received the green signal from the Americans, and in fact he visited the USA before the meeting. I met Lal Bahadur before he left as the head of the Indian delegation to Tashkent. He did have a history of heart disease, and had once, soon after assuming the Prime Ministership, been laid up for a few days on this account, and was nursed by his devoted and gracious wife Lalita Devi. He had been advised by his doctors to walk regularly every evening, and when he was too busy to do so, which was on most days, he would walk briskly up and down his office or in whichever room he happened to be. As Home Minister he did this several times in my personal office at Mansarovar in Delhi. But we all assumed that he was not in any serious danger of a recurrence.

The Tashkent meeting lasted several days. At one point we heard that the talks had broken down and that the Prime Minister was returning to India the next day. Then, suddenly, there was a report that a last-minute intervention had led to an unscheduled meeting between him and President Ayub, and that an agreement had been reached whereby the two sides were to pull back to the positions they held before the conflict started. This meant our having once again to give up the highly strategic Haji Pir bulge between Poonch and Uri, which would have so substantially improved our security position in the area, while of course getting back the Chhamb sector that we had lost. This dramatic development was soon followed by the shocking news that Lal Bahadur had died of a massive heart attack.

We were stunned by this unexpected and saddening development. Lal Bahadur, particularly during his tragically short period as Prime Minister, had endeared himself to the people with his mature and courageous leadership. Those of us who had the privilege of coming into closer contact with him will always remember his genuine simplicity and sincerity; his infinite capacity to listen to conflicting viewpoints and to find a seemingly impossible point of reconciliation between them; his unfailing courtesy and the deep consideration with which he treated his colleagues. It was also a tribute to his high sense of rectitude that, despite a lifetime in some of the highest offices of the land, he died with virtually no money or property to his name. It has always seemed to me that in some way his death was a sort of sacrifice for the cause of peace and friendship. I do not think the Tashkent agreement would have been so readily accepted by the country had it not been inextricably linked with Lal Bahadur's own personality and sealed by his death. Indeed, both India's wars took the toll of a Prime Minister; the Chinese War led to the passing away of Jawaharlal Nehru, and the Pakistani to that of Lal Bahadur Shastri. One was a defeat, the other a stalemate. It was left to Indira Gandhi seven years later to lead India to a victory unprecedented in its history.

Sadiq and I rushed to Delhi in an Air Force plane and were there for the arrival of the body from Moscow to pay our last homage to the departed leader. With the sudden departure of Lal Bahadur, the question of his successor immediately assumed paramount importance. Indira Gandhi had been included by the late Prime Minister as Minister for Information and Broadcasting and ranked number four in the Cabinet after Gulzarilal Nanda and T. T. Krishnamachari. I had kept close contact with her because of all the Congress leaders I found her to be ideologically and personally the most compatible. Also, because of my long association with Jawaharlal Nehru, Asha and I had developed cordial relations with the family. She had stayed with us in Karan Mahal along with Rajiv and Sanjay when they were quite small, and we had struck up an excellent personal rapport. I had succeeded her as President of the Bal Bhavan Society, and we worked together closely on the newly formed Jawaharlal Nehru Memorial Fund of which I became Secretary early in 1966. That she was not very happy in the

Shastri Cabinet was widely known. She had been included, of course, not only because she was Jawaharlal's daughter but also as a Congresswoman in her own right, having been President of that great organization for two years. But she was somewhat ill at ease in Parliament, and I suspect that her personal relations with Lal Bahadur were never really very warm. In fact it was in the air that after Lal Bahadur's return from Tashkent she may well have resigned from the Ministry. As it turned out, destiny had other plans.

Lal Bahadur was in some ways more conservative than she, and in foreign affairs also their perceptions regarding the two super-powers tended to differ. Because of this, Indira Gandhi was not unduly prominent in the Shastri government, although she did some valuable work behind the scenes when the Kashmir invasion took place. This led people who did not know her well to assume that she would not be able to step into her father's shoes. However, those of us who were close to her knew better. Her somewhat aloof mien concealed a passionate patriotism, a fierce sense of independence, a rare capacity to listen and an uncanny ability to gauge public opinion at any given point of time. Also, though she was and remained essentially a political person, her interests were by no means confined to the political arena. Women's and children's welfare, wildlife and the environment, preservation of our monumental heritage and the uplift of tribals, were among her many activities. Her somewhat unusual personal life since childhood had forged in her an inner strength that few people realized she possessed. It was only over the years that these capabilities became apparent.

Her election as leader of the Congress Parliamentary Party after Lal Bahadur's untimely death came as a tremendous shot in the arm for the younger generation of Congressmen. I was not formally a member of the party, because my office precluded this, but for all practical purposes I was among the younger group of post-Independence Indians who had great expectations from Indira Gandhi. I met her in her office immediately after her election, and as soon as she saw me she said with a smile 'I wish you were in Parliament now'. As I wrote to Dr Radhakrishnan in my next letter:

With the election of Shrimati Indira Gandhi as leader of the Congress Party in Parliament, and her subsequent appointment as Prime Minister, the torch of leadership has passed to one belonging to a new generation of Indians. The democratic and peaceful means whereby the succession was brought about, and the fact that the new leader is a woman of great political maturity and personal integrity, dedicated to modernity and progress, have raised fresh hopes particularly among the younger generation. We all wish her well in the extremely onerous task that she has undertaken.

In all the hectic events of 1965, an additional dimension was provided when our third child was born. During the war with Pakistan, Asha had been expecting a baby, and he was born in a Delhi nursing home on 6 February 1966. I have a special liking for classical names, so our first son being Vikramaditya (Vikram) we decided to call this one Ajatshatru (Ajay) after the great and wise king of Kashi who plays such a prominent role in the *Brihadaranyaka Upanishad*. My mother was very ill and was in Delhi for medical treatment. The birth of her second grandson gave her considerable satisfaction in the last few months when her health rapidly declined.

With Indira Gandhi's election as Prime Minister my determination to leave the State and enter national politics became overwhelming. I had been Head of State for over seventeen years, during which time I had completed my academic education, gained considerable experience of various aspects of national development, and travelled extensively abroad. I saw no good reason why I should spend the rest of my life cooped up in the State, when as a young Indian the whole nation lay before me to serve. The general elections were due in 1967, so I felt that I should resign well in advance. I wrote to the Prime Minister on 17 August 1966 as follows:

You will recall that over the last few months I have mentioned several times my desire to relinquish the office of the Governor and seek a more active participation in national politics. I have been Head of the State for almost half my life since I first became Regent in 1949, and I do feel that it will now be desirable for me to change my field of public activity, even though it will be a wrench for me to dissociate myself from close contact with the affairs of the State to which I belong. In view of the fact that the General Elections are approaching,

I write to seek your approval to my requesting the President to accept my resignation from the post of Governor of Jammu and Kashmir some time next month.

She replied promptly the next day:

Thank you for your letter of August 17. We have since met and talked about the matter raised by you.

You have mentioned your desire to be relieved of your present office more than once in the past and ever since you first did so, I have been giving anxious thought to your suggestion. I fully understand and appreciate that having served as Head of State in Jammu and Kashmir for nearly two decades now, you should wish to have a change and think of more active participation in national politics. With your ability and experience you would certainly be an asset to political life. You have held the office of Sadar-i-Riyasat and Governor with dignity and distinction in the midst of recurrent crises forced upon us largely by Pakistan's hostile and aggressive attitude. Your contribution towards the handling of these difficult problems has been invaluable.

However, as I have explained to you, we are passing through very critical times in economic and other fields. As the General Elections approach, our problems will increase. The threat to our borders also continues. In these circumstances and in this delicate situation I am worried at the thought of a new person taking up your important position. That is why I should like you to agree to stay on in your present office for the time being. I hope that you will find this possible.

I did not press my resignation, but my mind was made up. Meanwhile, I turned my attention to two matters which closely impinged upon the politics of the State. For the first time there were going to be direct elections from Jammu and Kashmir to the Lok Sabha, as previously the members were nominated by the State Assembly under our special dispensation. This meant that constituencies had to be carved out, and as six seats had already been allotted to us it was that number which now had to be delimited. Here again D. P. Dhar and I came into headlong collision. He had suggested that, on the basis of population, the seats should cut across regional boundaries; Ladakh should be with the Baramulla constituency, and part of Jammu's Doda district with the Anantnag constituency. To this I vehemently objected, because it would have blurred the regional identities of Jammu and Ladakh, and ensured perpetual Kashmiri domination over both these regions. My proposal, which I put forward strongly to the Delimitation Commission under Justice J. L. Kapur, was that because of its massive area which covered two-thirds of the whole State, and its distinct cultural identity, Ladakh should get one Lok Sabha seat. This would leave five, which could neatly be apportioned by giving three to the valley and two to the Jammu region. My view finally prevailed, and the pattern was adopted that continues till today.

The second point that I pressed was the importance of ensuring a fair deal for the three regions of the State. For this purpose I urged in my letter to the President dated 21 November 1966 the setting up of a high-powered Commission to look into the whole problem. After giving the historical background to the regional aspects of the State, I suggested that the terms of reference of the Commission might be as follows:

> The Commission will study the political, administrative and economic requirements of the Kashmir province, Jammu province and Ladakh district which together comprise the State of Jammu and Kashmir (excluding areas at present under foreign occupation). In particular, the Commission will review the developmental problems of these areas with a view to devising ways and means for their more rapid economic progress, and will make appropriate recommendations within six months.

In order to endow the Commission with some stature I suggested the name of the retired Chief Justice of India, B. P. Sinha, to head it, and also two other members, Subimal Dutt/Shankar Prasada, and B. H. Zaidi. As it turned out, these specific suggestions were not accepted, but the concept of a high-powered Commission to look into regional problems began to get accepted and finally resulted in the appointment, exactly a year later, of a Commission headed by the retired Chief Justice of India, P. B. Gajendragadkar, with Shankar Prasada and Badruddin Tyabji as members. Thus I had the satisfaction of setting in motion for the first time a process of ensuring a fair deal for the three regions of the State. It is a great pity that today, twenty years later, the process is still incomplete.

The Tashkent Agreement initiated the process of withdrawal by both armies from areas they had occupied during the war.

As the Pakistani troops withdrew from Chhamb–Jaurian, the long and painful process of resettling the refugees there began. It is not generally realized what a harrowing experience this constant uprooting must have been for them. One hears a great deal about refugees in other parts of the world, but there could hardly be a parallel to the travails of the Chhamb people. They were uprooted in 1947 and resettled with difficulty. They again had to flee in 1965 and were only resettled after great problems. And then, cruel irony, they finally lost their land again in the 1971 war despite our smashing victory.

The process of political mobilization for the general elections also began. Mir Qasim resigned from the government to take charge of party affairs, while Bakshi Ghulam Mohammad, despite the stigma of the Ayyengar Commission, showed startling political resilience and re-entered the political arena. In Jammu the Jana Sangh became active, especially in the towns and among the disgruntled refugees. They were sharply critical of the Tashkent Agreement which they attacked as a surrender to Pakistan. Meanwhile, despite the agreement, the ceaseless stream of propaganda from Pakistan-controlled radio stations in the occupied area, including a new one set up during the recent invasion called the 'Raiders Radio', continued unabated.

Increasing attention began to be paid to economic development in the State. I pressed for some major projects, particularly the extension of the railway line from Pathankot to Jammu. The old Sialkot–Jammu rail-link having been disrupted in 1947, the State was suffering great economic and administrative disability in the absence of a rail-link. Here again D. P. Dhar and I clashed. His proposal was that after Kathua, the first major town in the State on crossing the Ravi from Punjab, the line should be taken not to Jammu but to Udhampur. He sought to justify this on security considerations, but it was clear to me that this was a not so subtle attempt to denigrate Jammu, which remains the winter capital of the State for six months of the year. I opposed the suggestion and insisted that the line should first come to Jammu and then be extended to Udhampur, and this was finally accepted by the Government of India.

I had also moved to set up a federal structure in the University of Jammu and Kashmir with a separate, self-contained campus both in Jammu and Srinagar, so that academic growth in both the regions would be stimulated. The plan was drawn up by a Committee under the eminent economist Dr B. N. Ganguly, a gentle, soft-spoken man of great competence and integrity. This move, which was underpinned by suitable legislation, paved the way for a smooth transition to two separate Universities a few years later. For the last few years as Governor I took special interest in University affairs, and presided over meetings of the Syndicate. When Indira Gandhi became Prime Minister she resigned as Secretary of the Jawaharlal Nehru Memorial Fund and I was elected by the Trustees in her place. The administration of this Fund and its numerous schemes, including a major programme of Jawaharlal Nehru's publications and a Fellowship programme, was for many years a source of creative activity. The Banaras Hindu University Court in April re-elected me Chancellor for a further period of three years.

My kidney stone continued to give a lot of trouble, and the attacks of renal colic became more frequent. Asha and I went to London on a short visit towards the end of April for a medical opinion, which was that if the stone did not pass of its own within the next three months it should be operated. Finally, I had the stone surgically removed by the eminent urologist Dr D. K. Karanjawala, in the good old Breach Candy Nursing Home in Bombay which we had so often patronized in the past. It was a most painful and unpleasant affair, but happily since the operation the trouble has not recurred. Probably it was due to the water in Kashmir, which I soon gave up when I moved out of the State.

During the summer of 1966 an uneasy truce prevailed with Pakistan. Their agents were active, and on 16 May when Chief Minister Sadiq was on a visit to Baramulla and was being taken in a procession through the town a hand-grenade was thrown at him. Providentially he escaped unhurt, but three persons were killed in the attack and thirty others wounded. The 'Azad Kashmir' radio and the 'Raiders Radio' continued their campaign of vilification and communal poison. Within Pakistan itself Ayub Khan cracked down upon Sheikh Mujib in what was then East Pakistan for having put forward a six-point proposal for provincial autonomy, while Zulfikar Ali Bhutto made repeated statements regarding continuing confrontation with India. The link-up between Pakistan and China in the occupied territory of northern Kashmir was also an ominous sign, and it became clear to me that Pakistan would not rest content without one more confrontation with India. Pakistan having been born on the basis of hostility to India, the deepening of the hostility became almost an imperative for its successive rulers.

Meanwhile, Indira Gandhi began her first, somewhat uncertain tenure as Prime Minister. At that time C. Subramaniam and Asoka Mehta appeared to be her main advisers, or at least they came to be associated along with the Finance Minister Sachin Chaudhury with the devaluation of the rupee *vis-à-vis* the pound, a step which created quite a sensation at the time. In her broadcast on 12 June she explained this as being necessitated by the financial impact of the war and the drought, and said, 'The decision to devalue the rupee was not an easy one. It was taken after the most anxious and searching consideration.' Nonetheless, it was attacked strongly, especially by the communists who felt that it had been done under American pressure.

Several important Cabinet Ministers visited Kashmir that summer, including the Home Minister Gulzarilal Nanda, the Defence Minister Y. B. Chavan and Fakhruddin Ali Ahmed. While they all attempted to boost the Congress, now functioning under G. M. Sadiq and Mir Qasim, the redoubtable Bakshi Ghulam Mohammad also became very active. He called a major political convention in July, which was an impressive show of strength attended by thousands of delegates. This marked Bakshi's return to active politics after his ouster under the Kamaraj Plan three years earlier, and though Sadiq controlled the government it was clear that Bakshi was a shrewder and more effective political operator. The battle lines for the general elections thus began to be drawn, with Bakshi's National Conference emerging as the main challenge to the Congress in the valley, in view of the fact that the secessionist parties were banned, and the Jana Sangh providing the major challenge in Jammu.

Sheikh Abdullah was still in detention in the Kohinoor Bungalow (now renamed after him) at Kodaikanal in Tamil

Nadu. Several people visited him there, including the socialist leader Jayaprakash Narayan and Swami Ranganathananda of the Ramakrishna Mission. Conflicting reports emerged from the meetings. Meanwhile, there was an unfortunate controversy over the Dalai Lama's visit to Ladakh. When he proposed the visit the Ladakhi Buddhists, for whom he is virtually the Buddha incarnate, were overjoyed and began making elaborate arrangements to receive him. Sadiq and his colleagues, however, seemed to feel that this would be an unnecessary provocation to the Chinese, and wanted the visit postponed. This caused disappointment in Ladakh. Finally he made his visit at the end of August, and though it was blacked out in the national press it was a tremendous success and gave great joy and satisfaction to the people of Ladakh.

Indira Gandhi paid her first visit to the State as Prime Minister at the end of September. She was given a warm and affectionate welcome, and Sadiq and I accompanied her in a procession in an open car through the city of Srinagar. She visited several forward areas by helicopter and addressed troops there. I took the opportunity of discussing with her in detail the political situation in Jammu and Kashmir, and emphasized the importance of tackling the problem of regional disparities with courage and imagination. I reported to her on the detailed discussions in this regard that I had held with Sadiq, Qasim and D. P. Dhar, especially in the aftermath of my controversial statement and the reorganization of our neighbouring States of Punjab and Himachal.

In fact, soon after she left, the situation in Jammu suddenly deteriorated, with police firing upon college students resulting in three being killed on the spot and dozens injured. I drove down to Jammu and had a series of meetings with citizens, parents, students and officials. The whole city was seething with rage, and once again the weakness of a predominantly Kashmiri orientation to political and administrative authority was clearly shown up. Sadiq also arrived the next day, and was deeply shaken not only by the incidents themselves but by the tremendous hostility that they revealed between the people of Jammu and his government. We finally worked out a formula which involved setting up an Enquiry Commission under Justice B. D. Mukherjee, as well as other administrative measures, but I think Sadiq for the first time began to realize that my emphasis upon doing something concrete to meet Jammu aspirations was not merely rhetoric but based on real requirements. The only Minister from Jammu in the Sadiq Cabinet, Pandit Trilochan Dutta, resigned shortly thereafter at Sadiq's suggestion. It took us a full month to restore normalcy in Jammu and get the educational institutions, which had been closed after the police firing, reopened.

After prolonged hearings, the Delimitation Commission finally announced their decision. The Assembly seats were delimited so that the Kashmir valley got 42, the Jammu region 31 and Ladakh 2. As far as the Lok Sabha seats were concerned, my intense lobbying finally had the desired effect and they gave Ladakh one of the six seats. Three went to the valley and two to Jammu. This was a neat arrangement; the three Kashmir districts of Srinagar, Anantnag and Baramulla got a seat each, while in Jammu three districts—Jammu, Poonch and Rajouri —were grouped under the Jammu seat and the other three—Udhampur, Kathua and Doda—under the Udhampur seat.

Although I had agreed not to contest the elections immediately, we had an understanding that on one of the Jammu seats there would be a candidate of my choice on behalf of the ruling party, who, when the necessity arose, would resign his seat and make way for me to contest a bye-election. For this purpose I chose Brigadier Ghansara Singh, an elderly and highly respected ex-serviceman who had been Chief of Army Staff in my father's time. In 1947 he had been appointed Governor of Gilgit, where he was taken captive by the Pakistanis towards the end of the year and held as prisoner-of-war for many months. Subsequently I had appointed him President of the Dharmarth Council in my capacity as Sole Trustee of this family religious endowment. A man of great integrity, he decided that he would prefer to stand from Udhampur because of a large number of ex-servicemen in that area, and also because he was better equipped to handle rural rather than urban politics. In addition to him, the Congress chose the respected Head Lama of Ladakh, Kushak Bakula, for that seat, Inderjit Malhotra for the Jammu seat, and Shafi Qureshi, A. M. Tariq and Aga Syed Ahmed for the Anantnag, Srinagar and Baramulla seats respectively.

Elections were fixed for 21 February 1967 and excitement began to mount. Opposition parties, including Bakshi Ghulam Mohammad's National Conference, alleged misuse of government machinery, and there were violent clashes with the Congress in various parts of the valley. The major groups in the Plebiscite Front, claiming the blessings of the detained Sheikh Abdulláh and Mirza Afzal Beg, took the line that as the party was committed to the 'right of self-determination through a plebiscite', there was no question of its participating in the elections. A second group, supported by Maulana Masoodi and Mohiuddin Quarra, announced its intention to contest on independent tickets. When papers were scrutinized for the Legislative Assembly seats, 22 Congress nominees and one MP, Shafi Qureshi from Anantnag, were declared returned unopposed because the returning officer had rejected all the papers of opposition candidates. This was a blatant misuse of power, and it was a matter of regret for me that even Sadiq, who had come to office after accusing Bakshi of unfair tactics, should have had to resort to this. Of course technically it was the Deputy Commissioner of Anantnag who was responsible, but the development left a very bitter aftertaste. Opposition parties were furious; there were a series of angry protest meetings and processions. An opposition delegation consisting of Bakshi Ghulam Mohammad (National Conference), Pandit Prem Nath Dogra (Jana Sangh) and Dhan Raj Badgotra (PSP) went to Delhi to protest against the rejections. They demanded the dismissal of the Sadiq government and the imposition of President's rule to ensure fair elections.

I was also unhappy at this turn of events. Whereas I was certainly hoping that the Congress would win, I was totally opposed to the adoption of flagrantly unethical practices to ensure this. I said this very clearly to the Prime Minister when I met her in Delhi soon after the Republic Day early in 1967. I also reported to Dr Radhakrishnan, and, in my next letter to him, said: 'It was clear that the large-scale rejections had created a most unfavourable impression. On the one hand our foreign critics, and Pakistan in particular, fully exploited the episode to justify their contention that elections in Jammu and Kashmir have always been farcical, on the other indignation of the Opposition parties found considerable prominence and sympathy in the national press and there was widespread resentment within the State.'

It was decided that Election Commissioner Sundaram would visit Srinagar. He did so, but after his visit all he could say was that he had 'not found it possible to make anything more than a cursory scrutiny of the records and hearing the complaints made in person by party representatives', that he was 'satisfied that by and large the orders passed by the returning officers are justified from the records', and that 'the aggrieved candidates will have to resort to the usual remedy of filing election petitions'. He did recommend that the State Electoral Law should be brought into line with the national law as soon as possible so that the election petitions could be filed directly with the High Court as in the case of parliamentary constituencies. Despite some initial reluctance, I was able to persuade Sadiq to accept this and to issue an ordinance suitably amending the Jammu and Kashmir Representation of the People's Act. While this would not help the aggrieved candidates, it was a step towards trying to ensure fairer elections in the future and breaking away from the negative electoral syndrome that had dogged our footsteps ever since Sheikh Abdullah got 72 out of 75 of his members unopposed into the Constituent Assembly.

In this welter of activity, allegations and counter-allegations, I went off again on a visit to Varanasi where I presided over the historic Golden Jubilee Convocation, delivered my first Convocation address in Hindi, and received from the University the honorary degree of Doctor of Letters. I always cherished this Varanasi link and did not miss a single Convocation as long as I was associated with the University. On my return to Delhi I called on Indira Gandhi, who was recovering from an injury to her nose caused by a stone-throwing incident in Bhubaneswar. She was in high spirits, and was working hard to lead the Congress into the first parliamentary elections since the passing away of Jawaharlal Nehru.

India's fourth general elections were held on 21 February 1967. The Indian National Congress won a clear majority, but this was much thinner than in the third Lok Sabha. The conservative wing of the party led by Morarji Desai commanded over a third of the members of Parliament, and while Indira Gandhi won a clear victory over him in the parliamentary party elections for leadership, polling 355 against 169 for Morarji, it was clear that she would have to compromise with him if a stable government was to be formed. This she did, so that Morarji came in as Deputy Prime Minister and many persons from his group were adjusted at various levels in the Council of Ministers. As it turned out, the arrangement lasted only two years until the major Congress split in 1969, but these two years were vital for Indira Gandhi to firmly establish her national leadership and gather the strength necessary to take on the conservatives.

In Jammu and Kashmir, Shafi Qureshi was returned unopposed as papers of all opposition candidates had been rejected, while no papers were filed against Kushak Bakula in Ladakh. Both the Jammu seats were won comfortably by the Congress candidates, as was the Baramulla seat in the valley. The major upset was in Srinagar where, despite the entire strength of the Congress and the administration being pitted against him, Bakshi Ghulam Mohammad defeated A. M. Tariq and was elected to the Lok Sabha. In the Legislative Assembly the Congress won 26 of the 31 seats in Jammu, conceding three to the Jana Sangh and one to the National Conference. In the valley the National Conference was able to win seven seats, while two went to independent candidates including the ebullient young journalist Shamim Ahmed Shamim. Polling in Ladakh was scheduled only in June due to climatic conditions. Thus, overall, the Congress won 59 of the 73 Assembly seats for which elections were held, 22 of them unopposed.

I reported these results along with my analysis to the President in my last fortnightly letter to him dated 9 March 1967, my thirty-sixth birthday. With the general elections having greatly reduced the Congress margin, I felt that this was the time for me to enter the fray when Indira Gandhi needed all the support she could get. The political adrenalin had begun to flow, and I was not going to remain simply a passive spectator up in one corner of India. Asha and I went to Delhi on the 10th after putting Jyoti in the Lawrence School at Sanawar in Himachal Pradesh. I did not want to meet Indira Gandhi personally on this occasion as it could be faintly embarrassing; so I called over her Joint Secretary, Sushital Banerji, who had become and remained till his untimely death a close friend. He had first come to Kashmir a few years earlier as a member of Vishwanathan's administrative team on the holy relic episode. I asked him to convey to the Prime Minister that I was ready to resign and enter active politics now if she so desired.

On the 12th we were awoken in Mansarovar by an early morning telephone call from Sushital. 'The Prime Minister would like you to join her Cabinet as Minister for Tourism and Civil Aviation', he said. It was decided that the long-time Chief Justice of Jammu and Kashmir High Court, Wazir Janki Nath, would be appointed Acting Governor until a new appointment was made. On the 13 March, under Article 28(2) of the Constitution of Jammu and Kashmir, I submitted my resignation to Dr Radhakrishnan. In his acknowledgement of the same date he said:

> I accept your resignation of the office of Governor of Jammu and Kashmir and am recognizing the Chief Justice, Shri J. N. Wazir, as Acting Governor until a successor is appointed. As he will not be able to assume charge until the morning of the 15 March, your resignation may take effect from then.
>
> I know I voice the sentiments of the people of Jammu and Kashmir in particular and of India as a whole in thanking you for your valuable services to the nation.

On the morning of 16 March we assembled in the spectacular Ashoka Hall of Rashtrapati Bhavan for the oath-taking ceremony. Asha accompanied me, as did the wives of several other Ministers. There was a hush when Dr Radhakrishnan, dressed as usual in a white dhoti and south Indian turban, entered the hall to the fanfare of trumpets. One by one the Cabinet Ministers took the oath, Indira Gandhi being the first to do so and Morarji Desai the second. I was happy that I would be administered my first oath of office and secrecy under the Indian Constitution by a philosopher-President, although I was not yet a member of Parliament. Over the previous eighteen years I had administered oaths on numerous occasions to dozens of Ministers in Jammu and Kashmir. Now one major phase of my life had ended, and I was on the threshold of a new career. I had at last moved from the crown to the heart of India.

Index